"In his delightful new book, *Everything All at Once,* Bill Nye reminds us why he remains the high priest of nerd cool, spreading his irrepressible zeal for science and wonder across every page. Tracing his story from early years idolizing Carl Sagan and Steve Martin to the success of his Science Guy TV show and from recent stewardship of the Planetary Society to advocacy for climate change activism, Bill writes with clarity and vision, offering a compelling call to arms for scientifically literate citizens to *change the world.* As Bill puts it: 'Information is the currency of the nerd perspective,' and this book contains an embarrassment of riches."

—Heather Berlin, Ph.D., MPH, assistant professor of psychiatry, Icahn School of Medicine at Mount Sinai, and PBS and Discovery Channel host

"Our best teachers—in school and, later, in life—understand that learning is not accomplished by dragging students kicking and screaming to truth while we try desperately to stay afloat amid a sea of facts and figures. Nobody gets this like Bill Nye, America's national science-teacher-in-chief. In *Everything All at Once,* he smooths out the path to knowledge for us adults—all of us out here with families and jobs and attention span issues—like he did for us as children. He does it with plain truths, jolly good humor, and a respect for both his subject and his audience. If you don't like Bill, Bill, Bill, there's something wrong with you. And if you don't like this book, you're not trying."

—Michael Naidus, showrunner and executive producer
for *Bill Nye Saves the World*

EVERYTHING ALL AT ONCE

By

BILL NYE

Edited by

COREY S. POWELL

RODALE.

Copyright © 2017 by Bill Nye

Published in the United States by Rodale Books,
an imprint of the Crown Publishing Group,
a division of Penguin Random House LLC, New York.
rodalebooks.com

RODALE and the Plant colophon are registered trademarks
of Penguin Random House LLC.

Originally published in hardcover in the United States by Rodale,
an imprint of the Crown Publishing Group, a division
of Penguin Random House LLC, New York, in 2017.

Library of Congress Cataloging-in-Publication Data
is available upon request.

ISBN 978-1-63565-215-4
Ebook ISBN 978-1-62336-792-3

Printed in the United States of America

Book design by Ariana Abud
Illustrations by Bill Nye
Cover design by Amy C. King
Cover photograph by Jesse DeFlorio

10   9   8   7   6   5   4   3   2   1

First Paperback Edition

We can change the world. I'm sure of it.
Let's go.

CONTENTS

CONTENTS

PART III: HOW TO CHANGE THE WORLD

BILL'S GUIDE TO DOING EVERYTHING ALL AT ONCE

OBJECTIVE:
CHANGE the WORLD

EVERYONE you'll ever meet knows something you DON'T.

GOOD ENGINEERING invites right use.

Constraints provide OPPORTUNITIES.

Be part of the START.

Think COSMICALLY; act LOCALLY.

QUESTION before you BELIEVE.

CHANGE YOUR MIND when you need to.

Be OPTIMISTIC; be RESPONSIBLE; be PERSISTENT.

PART I

Principles of Nerd Living

CHAPTER 1
The Tao of Phi

This is a book about everything. It is about everything I know and about everything I think you should know, too.

I realize that may sound a little crazy, but I'm completely serious. We live in an age of unprecedented access to information. When you pick up your phone or open your laptop and go online, you are instantly connected to a trillion trillion bytes of data; that's a 1 followed by 24 zeros. Every year another billion trillion bytes of data move around the Internet, carrying everything from those important videos with kitty cats to the arcane but fantastic detailed results of subatomic particle collisions at the Large Hadron Collider. In that sense, talking about "everything" is easy. Everything you and I know, and everything we need to know, is already out there for the taking.

Yet despite all those whizzing ones and zeros—the collective intelligence of billions of human brains—I still feel that we seem awfully . . . well, stupid. We're not using all this shared wisdom to solve big problems. We're not facing up to climate change. We haven't figured out how to make clean, renewable, reliable energy available to everyone. Too many people die in avoidable auto accidents, succumb to curable diseases, do not get enough food and clean water, and still do not have access to the Internet's great busy beehive mind. Despite being more connected than ever before, we're not particularly generous toward, or understanding of, one another, preferring to hide behind denial and personal bias. The flood of information has effectively allowed us to know something about everything, but that *knowing* is clearly not enough. We need to be able to sort the facts and put our knowledge into action, and that is why I wrote this book.

I want to see humanity band together and change the world. I think it will take a special kind of personality to get this done: people who can handle the modern overflow of information, take in everything all at once, and select the parts that matter. It requires rigorous honesty about the nature of our problems. It requires creative irreverence in the search for solutions. The process of science and natural laws don't care about our politics or preconceptions. They merely set the boundaries of what is possible, defining the outer limits of what we can achieve—or not, should we shy away from the challenge.

Fortunately, there is a large and growing clan of people who think that way, who love nothing better than using the tools of reason to solve the most unsolvable-looking puzzles. We call them "nerds," and I humbly (proudly) count myself among them. I have spent a lifetime developing the nerd mindset and trying to master the admirable but often elusive qualities that come with it: persistence in the pursuit of a lofty goal, resilience to keep trying no matter what the obstacles are, humility for when one approach turns out to be a dead end, and the patience to examine the problem from every angle until a path forward becomes clear. If you already consider yourself one of us, then join me in doing more by applying your nerdiness to the big problems of the day, not just to trivia or minutia (although of course we will set aside plenty of time for those). And if you don't consider yourself a nerd yet, join me all the same: You will soon discover that everybody has an inner nerd waiting to be awoken by the right passion. My whole life has been a series of those kinds of awakenings, moments of epiphany when I became evermore aware of the joyous power of science, math, and engineering.

It happened to me with a jolt in the 11th grade in Washington, DC, when I took formal physics for the first time. In nerd culture, we might write that it was my phirst phormal physics, and we'd phind that phrasing rather phunny. The "ph" pronounced phonetically with the same fricative that produces the sound from the consonant "f" is from phi, the Greek letter φ. The Roman "p" looks vaguely like a Greek φ. In Greek, the "f" sounds a little breathy, so the Roman letter "h" serves to preserve that sound or tradition. I couldn't help myself—I had to stop typing and look up the roots of the "ph" in our words "physics" and "phosphate." When we see these "ph" words, we know they came to us from ancient Greek and then Latin. The scholars call it "transliteration," meaning "across the letters." Centuries ago a diligent, perhaps even enthusiastic, transliterator was inclined to add that "h" to the "p," and here we are. Phew.

This little digression encapsulates what it means to be so into a topic, so phocused and phascinated by some aspect of nature or the human experience, that people consider you—or more important, you consider yourself—a nerd. For me to really enjoy some deliberate misspelled wordplay, I had to think about the background of φ, "ph," and "f." I called on my knowledge that most English speakers pronounce the letter φ like the second syllable in "Wi-Fi," but Greek speakers pronounce φ like "fee," as in "Fee-fi-fo-fum, / I smell the blood of a nerdy one." And as I was checking that out, I recalled that φ has other intriguing connections to physics besides the linguistic one. It is the mathematical symbol denoting the golden ratio, a fundamental geometric proportion that appears widely in biology, economics, and especially art. In statistics, φ is a measure of the correlation between two separate factors, and so it is a crucial measure for distinguishing chance events from cause and effect in scientific experiments. Stick that in your back pocket.

You might regard the things I just told you as little more than bits of playful trivia, but I beg to differ. The knowledge I gained in my obsessive pursuit of φ changed me a little, and it just changed you, as well. The impulse to chase down details is central to the way I have solved problems throughout my life. It is also, not coincidentally, a defining nerd trait. Further evidence of my detail-oriented outlook: Long before the ubiquity of the Internet, my friends would say about me, "The party doesn't start until Bill gets out the dictionary." I like to know the background of words, the etymology, as well as the meanings of the words themselves. While I was reflecting on the digraph "ph" just now, I was also reflecting on what started that train of thought— namely physics, the study of nature, specifically energy and motion— and the joy I felt when I was first (or phirst) exposed to it.

The word "scientist" was coined in 1833 by the English natural philosopher William Whewell. Before then, the term was "natural philosopher," which sounds a little odd today but back then was a familiar expression. Philosophy is the study of knowledge; philosophers seek ways to know whether or not something is true, so natural philosophy was the study of what's true in nature. Or, in modern terms, a scientist is a natural philosopher seeking objective truths.

We look for laws of nature that enable anyone to make concrete predictions regarding the outcomes of tests and experiments. Science belongs to anyone who loves to think and look for connections in nature. It's not all about math and measurement, but wow, the math is what gives the predictions their precious precision. We can know the motions of distant worlds to such a degree of accuracy that we can land the *Curiosity* rover on Mars or send the *New Horizons* probe past Pluto with near-pinpoint accuracy. We can measure the exact age of a billion-year-old rock by clocking the decay of the radioactive atoms it contains. The combined power of math and science is amazing. That's why nerds are so drawn to them, but the insights that come from science are inspirational even to those who never plan to crunch the numbers.

Digressions and micro-obsessions seem to come naturally to those who catch the science bug. You may meet people who are into trivia or who adore the very small details of some area of study. It might be the names of counties in Maryland or Mississippi or the writing credits on *Star Trek*. My claim is that by learning those details, these nerdy people know something extra about a bigger picture, as well. The trivia expert (a trivialist?) has in his or her head a framework of an area of study, outfitted with the appropriate memory hooks on which to hang more information, which in turn enhances and fills in the bigger picture.

It's much easier to remember the names of counties if you have a mental picture of a state map that indicates what county is contiguous with what other county. The names, the map, the driving distances to the state capital—they're all much easier to keep track of if you have the details properly placed in a mental map. This big-picture frame, combined with detailed views of the world, enhances one's ability to navigate across county lines, carry out commands on a sailboat, or do a little brain surgery. I've found this to be a recurring theme in my most formative experiences: The details inform the big picture as much as the big picture organizes the details. Is it enough simply to know where Maryland is on a map? Maybe, but the extra knowledge of the state's inner workings creates a much clearer picture and brings with it immeasurable value.

I'm encouraged by how deeply nerd behaviors have been absorbed into mainstream culture. Not so long ago, the nerds were often defined in opposition to the popular kids in school. Nowadays, it's downright fashionable to exhibit an obsessive attention to detail—not just in science but in nearly anything. The kind of trivia knowledge that was once considered a TV game-show oddity is now a staple of Thursday-night social gatherings at hip neighborhood bars. Another encouraging sign: The most popular show on television as I'm writing this is *The Big Bang Theory*, which every week overwhelms sitcoms, dramas, and news commentary in the ratings. Tens of millions of people apparently identify with an odd bunch of characters who are nominally engaged in complex science but who also display their very human quirks.

Now, as much as I love nerds and nerd culture, I have also observed some worrisome trends over the last years that have motivated me to speak out. On the surface, things look promising. The increased attention to science, technology, engineering, and math (STEM) learning is terrific; it's great that programmers and tech-oriented entrepreneurs have become major celebrities in our culture and businesses. After all, our society is increasingly dependent on technology, and we will be in deep trouble if very few people understand the scientific ideas that the technology depends on. What's not to love? The growing fondness for trivia and geek-speak would seem like a great thing.

But the current pop version of nerd culture leaves me with a nagging concern. Geeking out—going fanatical about characters in comic books, for example—can be fun. It can develop a community of people whose lives are enriched by sharing a common interest. It brings together hundreds of thousands of people every year at Comic-Con and the like. But it's most definitely not the same as diligently studying math and science to grasp the complexities of climate, to engineer a disease- or pest-resistant crop, or to become one of our time-honored quintessential smart people: a rocket scientist. Geeking out is driven by the same instinct to hoard information, but the application of the knowledge is a different and considerably harder-working sort of thing. It takes me right back to my original thought: *Information* and *application* are very different things. When I talk reverently about the nerd mindset, I'm extolling the virtues of a worldview that involves gathering as much information as possible and being constantly on the lookout for ways to use it for the greater good.

The difference between information and application may seem obvious to someone reading along here, but it's not obvious to a great many people in the general public. There are charlatans and cult leaders out there who are able to pawn off false details and bias as fact and reason. I continually meet people who espouse their own stories of the origin of the universe and how we all got here, attempting to pawn off false information and bias as fact and reason. I'm not talking about traditional religious believers; I'm referring to people who string together some general concepts into their own quasi-physical theory of the big bang (or black holes, or some secret way to "fix" Einstein's relativity theory). I also meet and hear about too many people who exploit bits of scientific information to peddle worthless products, to promote counterfactual political arguments, to sow fear, and to justify sexist or racist thinking. In some cases, these people seem to genuinely think they are doing science, but they aren't. They may even think of themselves as nerds, but they really aren't. They adopt the language of physics or biology without having spent the time to know the established science and current thinking with regard to the stars and the space-time these stars inhabit.

Here's another important cautionary point: It's very easy for us, any of us, to draw faulty conclusions from a small sample of events. Superficial familiarity with nerd-style thinking may even encourage it. If you switch off the lights in your living room at the same moment two cars smack into each other outside, you might conclude that your switch-throwing caused a car crash. You might decide to never touch that switch again, at least not while there are any cars passing by. Or you might wait for a particularly disliked neighbor to cruise past and then start switching that switch as fast as you can.

In this extreme case, the cause-and-effect connection seems obviously wrong. I imagine the readers of this book have no trouble concluding that there is almost certainly no link between a light switch in your house and automobile-driver attentiveness (unless it causes the light to shine right into oncoming windshields, in which case, please switch it off right now). But what if it's a much more subtle effect, like reading that people who drink red wine are less likely to have heart attacks? Or that people with certain skin colors have lower IQs? At different times, some very influential researchers became convinced that such things were correlated. Should you believe them?

The geek-trivia attitude toward science is no help in sorting out false correlations. You might say that the connection between geeking out and doing science is itself a false correlation: Nerds adore geek trivia and make great use of it, but adoring geek trivia does not signify that you are thinking critically like a nerd scientist. The search for cause and effect—one of the foundational φs in physics—takes hard work and close attention. Even then, you have to be constantly on the lookout to make sure you aren't fooling yourself. There's no shortcut of just memorizing a few science buzzwords or adopting nerd-style hobbies. You're never going to change the world if you can't figure out the right ways to do it.

I had no idea that all this was about to hit me when I started in 10th grade at the exclusive Sidwell Friends School in Washington, DC. Until then I had attended the DC public schools, which had sharply degraded on account of the reckless management by the long-serving, drug-addicted mayor, Marion Barry. But after a kid was shot in a nearby DC junior high school, my parents had had enough. That's when they decided to send me to a private high school.

I had to really hustle to catch up with the other students and prove myself. After a year, though, I became comfortable in my new school. I got a circular slide rule as an award for being the school's "most improved" math student. It was a double validation to me. Not just that I could hold my own in a much tougher school, but also that I was surrounded by a culture that valued numbers and big-picture thinking and all the other nerdy things that meant so much to me then, and now.

Then it happened. I fell (phell?) in love with physics in Mr. Lang's class.

One afternoon back in the 11th grade, my friend Ken Severin and I set out to test for ourselves the equation for the period of a pendulum, the time required to complete each swing. If you're scoring along with us, the period is proportional to the square root of the length of the string, chain, or wire divided by the acceleration due to gravity. (If that didn't make sense to you, you can readily look it up. The pendulum equation is one more of the bazillion cool science things on the Internet.) We set up our own pendulum. It seemed to work okay, but the air drag and the friction in string fibers slowed this first one down too much for our liking. We rigged a longer string from the ceiling. Then we commandeered a stairwell and rigged a 4-story-long string with a sizable weight at the end. We were in business: The equation predicted the swing period of our pendulum with satisfying precision. It felt as though we had unlocked a mystery of the universe.

Science and what is nowadays called "critical thinking" take discipline and diligence. What enables us humans to make a living here on Earth is our ability to make predictions by finding patterns in nature and then to take advantage of them. Imagine how much easier it was to have a settlement or farm if you could count the days and know the seasons so that you planted and harvested crops at the optimal times. Imagine how much easier it was for ancient people to hunt successfully once they understood the migration patterns of game animals, their main source of protein. Imagine how much more fun it was for me and Ken Severin to watch the pendulum swing in exactly the way we'd predicted, because we were proving for ourselves the amazing thing about cause-and-effect scientific analysis: It really works!

As it happened, I was supposed to get a vaccination from my family doctor that afternoon. For over an hour, school officials paged Ken and me, but we didn't hear any of the public addresses. We were completely caught up in the experience, swept away with such nerdy ecstasy at each stroke of the stopwatch and slip of our slide rules—this being well before the days of electronic calculators, much less classroom computers. With each swing of the pendulum, we weren't discovering anything that many others before us hadn't already learned. That's the nature of being a student. But the process . . . the joy . . . Each detail of our experiment taught us a little bit more about sines, cosines, and tangents, about aerodynamic drag, about patience . . . There is nothing like it; science is empowering like no other human endeavor I know of.

The reason I was so rapt is twofold. First of all, that was my first true exposure to the predictive power of physics. In politics and all manner of social interactions, we use the expression "the pendulum will swing back," but we mean it in an impressionistic way: Attitudes will change again in the future, someday, sooner or later. In physics, the words have a precise, mathematical meaning. If you record your experimental setup accurately, you can determine exactly how high a pendulum will swing and precisely how far it will swing back. If you're as diligent as the 19th-century French physicist Leon Foucault, you can use a swinging pendulum to prove Earth is spinning, and even use it to determine how far you are from the equator. Push the equations a bit harder, and you can measure how our spinning planet distorts space and time around it, as NASA's *Gravity Probe B* satellite did—all because you took the time to understand a rock swinging on a string.

The second part of my passionate attraction to physics was the people. The boys and girl (only one at that time) in my 11th-grade physics class were my kind of people. It was an elective class, not required for graduation, and back then there weren't any special requirements to get in. We all just loved learning math and studying motion, finding out at the deepest level why things happen in the world. At this fancy private school, there were many very smart kids who had been raised by very smart parents with strong traditions of academic achievement. The kids in my physics class were especially brilliant. I swam as fast as I could to keep up, and I loved every moment of it. We made one another laugh with arcane, corny science puns, but that wasn't what really bonded us. We were all on the same team, you might say, all swimming toward the same goal: We wanted to get to the truth of how nature works, or at least as close to the truth as our minds could take us. In short, we were all nerds.

Thinking like a nerd is a lifelong journey, and I am inviting you here to take it with me. I truly believe it is the best use of your years on this planet. It keeps you constantly open to new ideas. Everyday events—tying your shoes, parking your car, watching a snowstorm—become revelatory experiments. When the results are not what you expected, you press to find out why and to figure out a better approach. That way of looking at the world soon becomes second nature. You will marvel that so many people around you don't do it. And I'll tell ya, if they did, we could change the world more quickly.

My brother reminds me of the moments when, as a kid, I stuck my hand out the passenger window of our car while it was moving at highway speed. I had seen the curved metal slats, or leading-edge flaps, come out of the front of the wings of airliners as they landed at Washington, DC's National Airport (now Reagan International Airport). I tried to shape my hand and move my fingers so that my hand became a wing and my thumb or index finger became a leading-edge slat. Well, I could get lift; my hand would get pushed up like a wing (still can). But my fingers are round; they're not strongly curved and thin like a slat or flap. A finger doesn't work as a so-called "high-lift" device. Nevertheless, I still routinely roll down the window and try my hand at flying. Knowing how wings generate lift proved extremely useful to me later when I worked as an engineer at Boeing, but even if I'd followed a totally different career, that type of insight would still have informed my life in a thousand other ways.

I'm always investigating nature, and people who spend time with me get used to it. I'm always looking for details that might lead me somewhere useful, whether I am plotting the future of space exploration in my day job at The Planetary Society or tinkering with a new solar collector on my home. Over the course of this book, I'll offer myself as a case study, sharing some of my most memorable nerd moments with you so that you can amplify your own nerdy tendencies and start changing the world. There is nothing more thrilling than the φ of physics, I've found, because it is the most powerful thing that humans have discovered. I make my way toward truth and happiness by embracing a point of view from which I can see the big picture.

I try to take in all the details—everything all at once—and then sift through them to find the meaningful patterns, as part of an effort to make the world a little bit better. But I'm just one guy. With you, and millions of others like you, we can turn our best ideas into action. We can solve our most pressing problems and live better lives. Join me and prepare to be amazed at how much lies within your power.

CHAPTER 2
Scout Lifeguarding

When he was a boy, my father was an outstanding Scout. He could hike for 20 miles a day by pacing himself carefully and picking the optimal paths. He could start a fire in the rain and then cook dinner over it. There are a few family pictures of my dad as a teenager decked out in a sharp Scout uniform, full of pride. He passed along a sense of mastery to my brother and me. We embraced the Boy Scout way of doing things outdoors as a way to learn about nature and to develop self-confidence in the sun, rain, and snow. If you can deal with adversity outdoors, we learned, you can deal with adversity in many other aspects of life. There was an implicit experimental method in all this, though I'm sure Ned Nye would not have put it in those words. If things go wrong while you're in the woods, you focus a bit and figure out what to do. If one kind of solution doesn't work, you try another.

Boy Scout training is a small, beautiful example of just how important a practical understanding of science and engineering is (and has been) for our success as a species. For instance, here's one important lesson I learned about keeping warm: Should you find yourself in a rainy, night-shivery situation, you need a basic understanding of how fire works in order to get one started. The trick is to split some firewood with your ax and use your knife to shave some thin strips from the insides of the logs; that way, you start with dry fuel. Have spare time while you're sitting around camp on a wet afternoon? Cut and carve yourself some "fuzz sticks" by peeling back thin strips along the length of sticks; aim for pieces about as big in diameter as a felt marker. Then keep them in storage for use at night. The thin, peeled-back strips of wood burn like crazy, and in turn ignite the stick, which you can use to light bigger, reasonably sized pieces of firewood.

These are the kinds of insights our human ancestors started figuring out perhaps a million years ago, probably in the region that is now East Africa. They surely did a lot of trial and error, though without the cool Scout pocketknives, and kept passing along what they had learned to the next generation—first by demonstration, much later by written instruction. They kept pushing into new territory, migrating out of Africa and into Europe, the Middle East, and Asia. With each push came new threats and new experiments into how best to survive the new environment. Our ancestors had to confront unfamiliar predators, discover what animals were easy to catch and what plants were good to eat, and develop clothing and shelter appropriate to the environment. They also had to learn how to work cooperatively. Their ever-growing knowledge was what allowed them to keep going . . . well, some of them at least.

I joined the Boy Scouts when I was 11 years old. Not too long after, I was helping an older boy named Robbie build a fire in the rain. We took turns carving strips from a log and stoking the fire. I'm not saying that we got carried away; I'm just saying, wow, you can get a fire really hot when you're motivated by being shivery cold on a damp afternoon as the Sun is going down. I remember our scoutmaster's comments: "Er, uh . . . wow, that's a pretty nice fire." (He meant, "Whoa, boys, that's a huge fire—way bigger than we need.") We were on a roll. Once the fire got going, the steam from the larger wet branches and logs became a reminder of how cold you can get when your clothes are wet while you're camping in the woods. Ever felt that kind of chill, followed by the relief that comes when the warmth hits you? It's one particular hands-on science experiment that you never forget.

The Scouts movement was started in England in 1907 by Robert Baden-Powell. He was a military commander who apparently got his inspiration while fighting colonial battles in Africa. He noticed that many of his troops were dying in the jungle, not from enemy actions but just by being lost on their own—and this was happening in a relatively warm region where food was growing on trees all around. In response, Baden-Powell wrote a book for his soldiers, a guide to the basics of wilderness exploration and survival. He later modified it and republished it as *Scouting for Boys*. It has sold 150 million copies and is, according to the *Guardian* newspaper, the fourth-most-popular book of the 20th century.

Having the knowledge and skills to survive in the woods is hugely empowering. The reality-TV show *Survivor* has versions in dozens of countries and has been a ratings success in the United States for more than 15 years, with plenty of spin-offs and other programming all loosely based on the idea that you can survive in just about any wilderness if you know what you're doing. As a Scout, I fully embraced the concept. You can do it. Follow the motto: "Be prepared."

Scout training lies at the far practical end of the nerd mindset.

People often raise questions about the applicability of ideas in math and science; many a parent has heard a child's laments along the lines of "When in life will I ever need to know the Pythagorean theorem?" Well, when we were Scouts learning about the combustion physics of wood or evaporative cooling in damp fabric, we knew exactly why these nerdy details are important. We didn't even realize, generally, that we were learning science. We just understood that these were the rules of the world and great things were possible if we learned to master them. That, in a nutshell, is where the whole adventure begins.

From before I can remember, my mother insisted that my brother, sister, and I learn to swim. As a kid suffering through the stupidly hot summers in Washington, DC (this was before our house had air-conditioning), I knew better than to miss a chance to jump into a cool pool. I became a natural swimmer, completely self-assured in the water. Maybe I couldn't make my hand fly by holding it out a car window, but I sure could use my arms to propel my body through the water. I could move up and down, side to side. I felt like I was flying down there, and in a certain scientific sense, I was. I wasn't worried about getting in trouble or drowning, because I was in my element.

Well before I was 10, the summers we spent at Lake Waullenpaupak in Pennsylvania got me confident enough that I wasn't concerned if the water was too deep for me to see the bottom. I swam down there with a mask and navigated through the clear dark-green lake, getting close enough that I could look at rocks and deep-dwelling fish. The deeper I dove, the cooler the water felt. Looking back, I realize those watery adventures were intensifying the scientific impulse in my brain. The fish seemed to take little interest in me; they had places to go and mates to seek. I felt like I was seeing nature almost as if other humans did not exist. I also experimented with the effects of buoyancy and resistance.

In high school, I took and passed the Senior Lifesaving test. This is a class in which you nominally learn how to rescue a drowning victim. It takes another, more intense form of concentration to rescue a classmate who is only pretending to drown. You swim out to the fake victim and pike-dive in front of him (or her). From his point of view, you disappear. Underwater, holding your breath, you twist him around by the knees so that his back is toward the edge of the pool or shore. Then you sling your arm over his chest and sidestroke your way to dry concrete or sand. It wasn't just theoretical. We got drilled on it over and over. I admit it took tremendous effort to rescue a certain classmate victim who happened to be a young woman, who happened to look amazing in a bikini—but I managed.

Next I became a Scout Lifeguard, putting my knowledge to practical use. Becoming a Scout Lifeguard is very much like getting your Senior Lifesaving patch, but Scouts are also expected to learn to row because so many camps feature swimming areas marked out in lakes. In those days, the Scout rescue procedure was a little different from the one I learned in high school. You were expected to dive completely under the drowning victim; that, apparently, was the Scout style. Then you came at the drowner from the seaward or farther-from-the-dock side. After securing the victim, you had to swim in a 180-degree turn to get that guy (no women around at Scout camp) and you headed back toward shore.

The idea was to simulate a real-world situation. Knowing your technique was not enough. You had to understand human nature and know what to do when the drowner's instinctive reactions were counterproductive, even dangerous, to the rescuer. In either case, Senior Lifesaving or Scout Lifeguard, the main challenge of the final test is that the victim is supposed to be panicked—not stricken with panic so much as violent with it. We all looked forward to playing the part of the victim. It was a chance to thrash; you even might get the opportunity to legally punch an acquaintance or a rival in the face with a well-placed arm flail. The final exam became one type of rather tame opportunity in a coed high school pool and a whole other, more violent affair in the much larger, all-boys Scout swimming area.

By long tradition, the Scouts seeking the Scout Lifeguard badge had to "rescue" the camp counselors, who were a few years older, bigger, stronger, and ornerier than we were. It was daunting. How was I, skinny young Bill, supposed to drag in Big, Strong Counselor Man? I had the training and the knowledge; the X factors were courage and commitment. All of us who sought to pass our final swim test were lined up on the dock, and we were accompanied by the camp counselors. The drowning-actors swam out about 25 yards from the dock, and on a leader's signal, those strong young guys pretended to become panic-stricken, big-splashing victims. As you might imagine, they were having great fun becoming as charming as an angry bull and as easy to wrangle as a greasy anvil. Probably because of my precocious (that is, obnoxious wisecracking) tendencies, I got assigned to "rescue" the counselor whom everyone called Big John. I was 15; he was 19. He was a good 14 inches taller and 50 pounds of muscle heavier. Big John was determined not to let me get an arm across his body and swim him in to the dock. I was just as determined to show that my training and knowledge could overcome his desire to mess things up.

They tell you over and over that if your drowning victim is being wild, just roll with it. Once you've got your victim in hand or in arm, let him (or her) flail. If he dips his head in the water, let him. We were assured that he would soon flail himself again to get his face out of the water, and while he was catching his breath, you could catch a break. Then you could begin again your journey toward the shore. As reasonable as that might sound, Big John was a full-on flailing machine. I looped my arm across his chest and started for shore, pulling hard with my free hand and kicking just as hard with what is called the "inverted scissor kick." It's an unnatural position, and it takes practice, even if you don't have a Big John fighting you every step of the way.

They also tell you that if your victim is too thrashy, too violent, or, as in this case, too obnoxious, you have to secure him with two arms. You have to hold him from above and below. Big John was able to twist his way out of every one-arm hold I tried. So after a few failed attempts at that method, I tried holding him by clamping two arms around him. This meant the only propulsion I had was by means of my inverted scissor. It took a long time to drag flailing John to the dock by those means—yet I succeeded. Very much to my surprise, I was the only guy that morning to get his victim to the dock.

I absolutely do not attribute my success to any superior athletic ability. Much bigger guys were dealing with much less motivated counselors that day. Instead, I think what helped me was my approach to the problem. The Scout instructors had told us what to do in each situation. I had a precise rule book to work from. You have to swim under the guy and surface beyond him. You have to swim in a semicircle. You have to invert your kick. If he's acting wild, you have to hang on with both hands. I was in the internal reality of the rescue simulation; I wasn't considering my counselor's motives for making the task so difficult, but I accepted the terms of the situation and was solely focused on finding a resolution. I had to just get 'er done. So I did.

I am pretty sure the only reason the other guys didn't get their counselor victims to the dock was because they knew they didn't have to. Everyone there could swim. Everyone there had spent some time in the water learning the inverted scissor. Everyone was nominally strong enough to do the right thing and get a kid or a big grown-up to shore in a real, serious situation. After each Scout had shown his skills to the counselor in the water, even if he didn't get his "victim" to shore, the counselor said, "Okay, fine, you passed. Let's get to the dock." But I was driven by a bigger goal. I wanted to do it for *really* real. I wanted to put the theory to a hard test—I did, and it turned out that it worked. Well, it worked for me. The other guys were all standing on the dock, arms akimbo, unimpressed. Their attitude toward me was, roughly, "Are you quite done? The rest of us passed without all the extra splashing."

I've often thought about that morning in the many years since. I took everything I knew and, in a fit of nerdy ambition, tried to do something slightly more ambitious than what I thought was possible. We've all had that feeling at one time or another. It happens when you first learn to ride a bicycle, or master a gymnastics move, or find yourself running to second base after hitting a double, or perform a piece of music flawlessly for the first time. It's also the feeling of the scientist running experiment after experiment until the data begin to fall into place and a deeper awareness emerges. You can surprise yourself, I realized, if you focus, follow the procedure, and stick with it, stick with it, stick with it.

Not that I was spinning any such highfalutin notions at the time. All I knew was that I had studied all the rules of lifesaving and lifeguarding, and I'd be damned if I wasn't going to see them through, Big John or no Big John. My motivation and belief were what convinced me that this task, even though it looked nearly impossible, could be accomplished if I trusted myself.

It was what I believe we call a "life lesson."

CHAPTER 3

Me Against the Rock

Nothing sharpens your appreciation for science and engineering like a nice, rousing life-or-death situation. But hold on—I'm getting ahead of myself. Let me back up and take you on a return trip to my childhood in the 1960s.

As you have probably realized by now, I love being in the water. I also love spending time *on* the water, navigating under my own power. I trace that second love to age 11 when, along with some other tenderfoot Scouts, I got in a canoe for the first time. My scoutmaster, "Uncle Bob" Hansen, was a stockbroker, a gentleman farmer, and a consummate outdoorsman. Significantly, he also had a close friend named John Berry, a champion canoeist who built his own fiberglass boats at home. Listening to our scoutmaster talk, you'd think this guy practically invented the decked canoe. A decked canoe looks like a kayak but with subtle differences. A canoe has more bottom; its hull is rounder than that of a kayak, and its bow and stern curve up toward the sky a bit more than a kayak's hull does. If swimming taught me respect for the physics of water, my encounters with the canoe made me appreciate that science alone is not enough. When you are in a dangerous spot on a river, engineering is awfully important, as well.

Both the kayak and the canoe have a long engineering history behind them; they are products of different cultures working out different solutions on different continents. Kayaks were invented by the Inuit, Yup'ik, and Aleut peoples of North America. The earliest known canoes were built in northern Europe, though the same basic design appeared (independently, apparently) in Australia and the Americas. The similarities are no coincidence. Everywhere, people were trying to solve the shared problems of getting food by navigating across the water. On first glance, a canoe looks like a kayak, and a kayak looks like a canoe. Inspect them for a few more moments, though, and you can see that river people and ice-fishing people made distinct choices to optimize their various vessels' performance. For carrying loads like animal pelts or sacks of rice, a flatter-bottomed canoe has more room and is more stable. For chasing and hooking fish, a kayak is more maneuverable, especially perhaps around small floes of ice.

Each type of boat requires its own paddle and its own technique. A kayaker wields a long paddle with two blades, one on each end. By tradition, and by two or three dozen millennia of trial and error, a canoeist goes forth with a shorter single-bladed paddle. In a kayak, you're sitting down. Your legs react and support the force of each paddle stroke, but sitting down restricts how much leg you can put into each pull of your arms. In a canoe, I quickly learned that you've got to kneel, not sit, and you'd better keep paddling or the river will do whatever it wants with you. Your hips and thighs can provide a good bit more force to drive and steer the boat, so much more that it's hard to muster enough force with your arms to compensate for all that, unless you put both hands in the service of a single paddle blade.

The 10 or so canoes issued to us Scouts that summer in 1967 were open boats, sturdy, time-tested, and made of hard-to-damage aluminum. My fellow Scouts and I found that we could run them over the river rocks pretty recklessly. Afterward, the boats would show scars, but they'd still be quite river-worthy. Our teachers had shown us how to angle the keel and use the river current to ferry right or left, how to figure which rocks were deep enough to slide over and which would stop one end of the boat and spin you around. They had shown us nominally how to shoot white-water rapids. Nevertheless, we—I'm pretty sure it was all of us, not just me—were often terrified. We were not yet fully in control of our nerd powers, and we knew it.

There are a couple of tricks with a single-bladed paddle that enable you to push a lot of water around in a hurry. You're probably familiar with Newton's third law of motion, whether or not you call it by that name: For every action there is an equal and opposite reaction. That's the scientific principle that sends rockets into space, with the exhaust throwing the mass of the fuel down while the mass of the rocket soars up. In the same way, when you push on the water, the water pushes back and moves the boat in the other direction. When you pull on the water, you pull the boat toward your paddle. When you really get smooth with it, moving the boat feels like magic. But as I am so fond of pointing out, it's not magic; it's *science*. It is perfectly, beautifully predictable—if you know what you're doing.

When you are in a canoe, you don't need to study physics to understand action and reaction. Action and reaction, viscous drag, wind resistance, turbulent flow, force equals mass times acceleration . . . you don't have to know the theory behind these things to master a kayak, but you sure do need to know how they all work. You learn it somatically—in virtually every fiber of your body—when you put paddle to water. That's what the early Inuit, Aboriginal Australian, and other water-faring cultures did, and that's what I was doing all over again on Pennsylvania's run of the Youghiogheny River in the summer of 1967. I was mastering a fundamental piece of technology and learning how nature works, just as other eager kids had done for thousands of years before me. If you've ever gone paddling, you know exactly what I mean. And if you haven't, well, I highly recommend it.

A lot of this is intuitive, but some important ideas are not.

For the hard-core kayaker or C-1 decked-canoe paddler, a major rite of passage is the "Eskimo roll." Here, the physics is predictable if you know what you are doing. It is utterly unforgiving if you do not. Skilled kayakers from any culture can roll their boats completely over, under, around, and back upright again, all in a single fluid (pun intended) motion. Their head and torso go completely underwater, but only for a few moments. This is where more engineering comes into play. Whether you are rolling for fun or desperately trying to recover from capsizing, when you attempt an Eskimo roll, it's all too possible to get stuck head-down with no source of air. That's bad. Then you have to either abandon your boat by wrestling free and swimming straight down—no easy task in a close-fitting kayak or decked canoe— or stroke with just the right motion, with the paddle pulled hard from behind your head toward your thighbone, to twist yourself back upright. In other words, you can easily drown beneath your own, nominally very maneuverable boat.

This buddy of my scoutmaster, our stalwart Mr. Berry, was accomplished on the water, almost ridiculously so. He smoked a pipe while he plied the white water; that's how cool and calm he was. One chilly morning in a calm part of the river, while showing off his canoe-paddling techniques, he performed an Eskimo roll and spun down and back up so swiftly that his pipe didn't go out. He puffed it right back to glowing orange life. I realize now that this may have been an illusion. Maybe the tiny fire in his pipe did go out, and what I recall witnessing was just steam rising from the hot tobacco leaves. Whatever actually happened, though, I've never forgotten the feeling that the sight inspired in my young brain. Mr. Berry had such command over his boat and its position on the river. He was not the slightest bit concerned about oncoming rapids or rocks, let alone going upside down in a river eddy. He understood the exact engineering capabilities of his canoe, and he had his somatic knowledge of physics down pat. I hoped to be that good someday: not to worry about what might go wrong, and be ready to recover in case something did. It's a confidence that gets justified only after you prove to yourself that you can consistently paddle and roll with the river, no matter what challenge might come along.

Steering a canoe is a matter of combining well-timed quick flicks by the guy in the stern with a few real hard paddle strokes by the guy up front. Quickness is key. With the occasional missed opportunity in bow pulling and stern pushing, now and then one of the boats would capsize. It's a little bit like accidentally driving your car into a tree. I don't recommend it. Although the consequences of a canoe capsizing were generally not as serious as a car crash, if you did capsize, you were cold and embarrassed. More than once, I saw my fellow Scouts have to crawl to shore or to a high and dry big rock, turn the canoe upside down to get the water out, and then recover their now-wet camping gear.

Along with providing much of the power, the bowman is also the first guy at the scene of a crash, if you catch my drift. As a young Scout,

I was put in the bow, and I often paddled for all I was worth, exhausting myself by day's end. As much fun as I was having, I was also constantly aware of the consequences of any small error in judgment. I knew what it was to tip over, to capsize, to get everything you have with you for the weekend instantly soaking wet and cold.

Then one time it happened to me: We hit a rock, and over we went. There were the standard two of us in our canoe. I claim to this day that it wasn't my fault. I was up front, pulling and pushing. The guy in the back was supposed to do the steering (right?). In any case, when things go wrong in white water, they go wrong very quickly. This is true of any high-speed activity that depends crucially on human reaction time—including skiing and, more commonly, highway driving. We smashed the boat on a rock. With the bow dead stopped, the stern caught the current, and the whole boat spun around. Water came over the gunwales (sides), and all our gear went flying. My knees, shins, and feet were soaked. We easily could have found ourselves in a part of the river from which there was no simple exit, but we managed to maneuver the boat near enough to a big rock that we could crawl out onto it.

By that point, our canoe had taken on so much water that it was behaving more like a full bucket than a boat. Once we were secure on the rock, we tipped the canoe up, rolling it on its side so it could drain. It wasn't the life-or-death kind of event I referred to at the beginning of the chapter, but it was humiliating. When we finally got back to camp, it seemed like most of the Scout troop was waiting for me. Along with the embarrassment, I had a desire. I wanted to do better—eventually. In the moment, I was so frustrated that I never wanted to do anything like that again. Then, after a few hours and days of reflection, I really wanted to do it, to get good at it, to get confident. This is a common process. It's part of being a nerd. It goes with the old saying, "No one cares about you when you're down. It's how you get back up that they'll notice." I wanted to get back up from this little wreck and become a great, or at least good, paddler.

Years and canoeing seasons went by. Four of them, to be precise. Although Uncle Bob and the other grown-ups moved our Scout group around on a few different eastern rivers, this day we were back on the Youghiogheny in Pennsylvania's Allegheny Mountains. Now I was 15 and had graduated to the stern of the canoe, and I had a young guy named Ken in the bowman position up front. Things were going well. The water was rough but not too rough, the day was clear, and there was no wind to speak of. Just a few minutes earlier that morning, Ken and I had witnessed another boat run aground, get spun sideways, and capsize. He was spooked by it. I knew that feeling and shared it that day, although for different reasons: Ken didn't yet feel the science and engineering of the canoe in his bones, and that kind of uncertainty breeds fear. He hadn't had the chance to be confident.

At first, things were smooth, and then very quickly they weren't, in the sudden way that can happen on a river. The current had picked up, and I realized, a moment later than I should have, that we were headed for a medium-size rock. This was not one of those monolith-of-death-looking rocks, which are out there. No, this rock was big, but smaller than our canoe. Still, we had some speed up, and I knew that if we hit it—well, we'd be in the middle of the cold, swift current with no easy means to recover. At that moment, I saw Ken recoil and freeze. The rock was, from his point of view, going to kill us. Four years earlier, I might have frozen, too. Even now, I clearly remember my thinking the moment when I saw Ken's mounting panic and could visualize our impending doom: "Wait, I'm the senior guy here. I can handle this." All that nerdy knowledge in my head had transformed from information to action. My reaction was intuitive, automatic.

Without thinking—absolutely without thinking about it—I shouted,

"Brace!" Ken knew the command and automatically braced his paddle across the gunwales of the boat so the paddle was no longer controlling our direction. Meanwhile, I forcefully steered us around the rock, just barely missing it. I was in real danger there on the Youghiogheny River, and I really avoided it. I took control. In that moment, I had 4 years of canoeing under my belt. I instantly assessed the problem of the oncoming rock, the design of the canoe, the limited abilities of my guy in the bow, the dynamics of the water, and Newton's third law. My bowman was a guy just like me, only a few years younger. He probably had all the physics he needed, too, but what mattered was confidence bred by repetition. He didn't know how to put his knowledge into action.

There were going to be unknowns, but I knew how to roll—or flow—with them. It wasn't just a way of thinking about paddling; it was a way of thinking about the world. It was a way of living. There are always going to be unknowns, and there are always going to be moments when I (or you) look into a crisis and then . . . either panic and freeze or realize "I know how to handle this." In the few seconds before we hit the rock, I understood with crystal clarity what I could accomplish by scanning the full database in my brain and putting to use what I knew. By paying attention to everything all at once, I was able to execute a solution to the problem of the rock.

It may sound like I'm making a big deal out of a small rock in a beautiful river replete with rocks, but it was a seminal experience. I was coming of age and realizing that I could handle situations, dangerous ones, if I paid attention and kept my head in the game. That moment changed me for the better. If we had hit that rock, I'll readily admit, we wouldn't have died and likely wouldn't have been injured in any lasting way. We would've gotten dunked, and possibly stranded. The adults may have been faced with the sorry task of paddling hard upstream against the current to recover our gear. Maybe one or both of us would have gotten banged up a bit on the rocks. We would have been cold and miserable for the rest of the day. Perhaps more important, the other Scouts would have made big fun of us. We would have been uncomfortable had we capsized, but that was the path not taken. Instead, we dodged the rock and carried on.

That day with the rock made me want to use good judgment about . . . everything, really. Is such a thing possible? Is it even desirable? After all, going wild now and then might not be all bad. But preparation for anything always pays off. What's wonderful about nerd knowledge is that it is there if and when you need it. So when I think about the changes I want to see in society, I think about avoiding the rock in the river by being prepared and in control. I think about the filtered knowledge, applied just the right way. I feel that the canoe, the rock, energy policy, and climate change are connected. No kidding.

I know many people who would move past such an experience and forget about it. I know others who would go the other way and interpret it as a spiritual message. For me, it was unforgettable, but it was not evidence of a higher power looking out for me. What saved Ken and me was nothing more or less than the steady training by the older Scouts and the patient guidance of Mr. Berry and Uncle Bob, the scoutmaster, along with his assistants. They had taught me about the river and about the design of the boat and paddles. They had drilled the necessary information into me. They had trained me in a way that didn't lead just to theoretical knowledge but also to muscle memory and preparation that were completely related to the real-world experience of many different possible scenarios. When I was actually faced with one of those situations, as the saying goes, "the training kicked in."

There's an art to having a productive scientific epiphany. Life delivers plenty of instructive experiences to all of us, but it's up to us to figure out what to do with them. My recommendation is to internalize them as a real-world practice and not just as information on a page. We miss the opportunity way too often. I try to pay attention to everything around me but then apply a strong mental filter so I can focus on the things that really matter. I pay special attention to events that reveal new details about how the world around me works and about how I can make it work for us.

This is a stark difference between the religious and scientific viewpoints. If you depend on miracles to make great things happen, you rejoice in the moments when control is taken out of your hands. If you think like a nerd, you celebrate the moments when you are most *in* control—when you see a theoretical practice play out in a real way, in real time. It's not that you take life's marvels for granted; it's that you work hard to understand them, to learn new things from them, and to add them to the storehouse of information in your brain. The more you learn, the more there is to enjoy about the world, and the more you can do to take control of it.

CHAPTER 4

When Slide Rules Ruled

A long with his many other remarkable contributions to the world of music, Sam Cooke recorded one of the greatest singles of the 1960s, the song "Wonderful World." It's a soulful pop song set to a cha-cha rhythm, at about 130 beats per minute. Sam sings:

> *Don't know much about geography*
> *Don't know much trigonometry*
> *Don't know much about algebra*
> *Don't know what a slide rule is for*
> *But I do know one and one is two*
> *And if this one could be with you*
> *What a wonderful world this would be*

People of all ages love this song, not only for the memorable melody but also for the very relatable lyrics. I loved it as a kid, and I still do.

Like Sam, I often feel I don't know much about geography. (I'm pretty sure I could not locate Alexandria, Massilia, Syracuse, Antioch, Gades, Argo, and Carthage on the Mediterranean Sea's coasts, as was required on the Cornell University admission examination in 1891. Although, come to think of it, I could find Alexandria.) However, I absolutely do know what a slide rule is for. I came of age with one of them hanging from my belt. If you want to understand where modern nerd culture came from—how we learned to see numbers and information everywhere we look—then consider learning about the slide rule. Swimming and paddling around gave me a feel for how science works, but slide rules introduced me to a lot of the subtleties of how to use the method of science for real. When you can decode the laws of nature through numbers . . . well, no offense to Sam Cooke, but that makes for a genuinely wonderful world.

A slide rule is a calculator, but it is not, absolutely not, electronic. It is just a set of beautifully machined wooden, plastic, or metal strips that have detailed sets of rule markings on them and that can slide past each other. Hence, *slide rule*. To understand how you can multiply and divide and find squares of numbers, square roots, cubes, cube roots, and several useful trigonometric functions in a flash using nothing more than a couple of sticks, start with this: Have you ever used a piece of paper or cardboard to figure out how wide something is? You put a pencil mark on the paper to note the dimension, then hold the paper against a ruler to read out the width. If a single piece of paper isn't large enough, you put a second sheet next to it and put the mark on that second sheet instead, knowing you'll add that partial dimension to the full dimension of the first sheet. Simple enough.

If you're a more numbers-oriented sort, have you ever measured a length using two rulers? With two rulers and a little thought, you can add the full length of one ruler to the partial length of another. Here we go. A side table or an end table next to your bed might be 16 inches (40 centimeters) wide, for example. Set one 12-inch (30-centimeter) ruler on the table, and then set a second ruler next to it. Read that second ruler's number. I'm confident you can find the 4-inch (10-centimeter) mark. Add 4 to 12 and you generally get 16 (inches). Add 10 to 30 and you'll get 40 (centimeters). This approach works fine for round or even fractional numbers, with just a little adding-in-your-head work. (From here on out, I am going to use metric units. They are the very essence of the everything-all-at-once universal data conversion system. Our hands nominally have 10 fingers, so we use 10 digits, 0 through 9, with which we express any number in nature.)

A slide rule does the same basic thing, but it adds and subtracts using specialized scales marked on lengths of bamboo, magnesium, plastic, or even ivory. There are generally two main pieces: the "slide" and the "body." Both have scales precisely engraved or printed on them, but instead of demarcating regular lengths, those scales are marked to indicate a number's logarithm. By the way, slide rules are fitted with a crosshair—a thin wire or length of real hair embedded in plastic or glass to keep the all-important numbers on the slide scale and the body properly lined up. Gentle reader, do you know what that crosshair carriage is called? The "cursor." This term was coined centuries before computers had cursors. People everywhere use the word today without knowing that it emerged from the early history of my fellow nerds. That secret connection fills me with pride.

If you don't happen to remember logarithms, really, there's nothing especially complicated or terrifying about them. They go like this: 100 is 10 squared, or 10^2. The logarithm of 100 is just 2. The number 2 is written above and to the right, and it's called an "exponent," from the Latin for "placed away." The exponent describes the logarithm. The number 1,000 is the same as 10^3, and the logarithm of 10^3 is 3. If you multiply 100 times 1,000, you get 100,000. I suspect you already got that. The number 100,000 is 10^5, which is another way of saying that the logarithm of 100,000 is 5. It's lovely. It's 10^2 times 10^3, which equals 10^5. You don't need to multiply the numbers. Just add the logarithms (2 + 3 = 5). Logarithms make life easier in a nerd-loving way. *Exponentially* easier, you might say. And there's more, much more: You can have logarithms that are in between the round numbers. The logarithm of 10 is 1 (10^1 = 10), and the logarithm of 100 is 2. You can intuitively see that the logarithm of 50 is going to have to be between 1 and 2. In fact, it's very nearly 1.70. Now here comes another dose of numerical beauty. The logarithm of 1 is zero (10^0 = 1), so the logarithm of 5 is very nearly 0.70; it's just the logarithm of 50 minus the logarithm of 10. Can't believe a number raised to the zeroth power is 1? It is, and I'll prove it to you in one sentence: Anything multiplied by 1 is just that same anything, so anything raised to the zeroth power has to be 1 to make the multiplication work. Cue the spooky music.

Logarithms are an essential part of the language of science because they provide a convenient way to write down the staggeringly large and small quantities you encounter when you stretch beyond human-scale perceptions. How many stars are there in the observable universe? Oh, about 10^{23}. How many atoms are there on Earth? About 10^{50}. Logarithms are what make slide rules so delightful to use, too, once you get familiar with them. When you slide one logarithmic scale alongside another logarithmic scale and read the added-up sum, the way you did when we were measuring the end table, you are no longer getting the arithmetic or 2 + 2 kinda number. You are now getting the added-up logarithms. In other words, we are multiplying by adding. When we move the slide the other way, we are dividing numbers by subtracting logarithms. Whoa . . . Cue more spooky music.

When I was in high school and college, we used to have races to see who could multiply, divide, multiply by the number π (pi), and then find the square root of the resulting number the fastest, or similar festivities. It was a standard competitive sport among the nerds. I was pretty good at it. But Ken Severin was the major league. He got a perfect 800 on the SAT Level 2, back when that was the highest score you could achieve. He went on to the California Institute of Technology (Caltech) and became an expert on using electrons to take pictures of tiny things, with a now-standard tool called a scanning electron microscope. Later, as Dr. Severin, he joined the faculty of the University of Alaska and established its Advanced Instrumentation Lab for geology. He was my best high school buddy. We had nerd adventures together, tinkering with resistors, transistors, capacitors, and the like.

Don't worry if the specific details of the slide rule still sound a mite confusing. Learning to keep all the numerical patterns in your head takes practice, and that's kind of the point. Mastering math, science, and any other advanced skill is hard work. A slide rule was therefore a badge of intellectual pride. No, more than just a stinkin' badge. It was a giant airport runway beacon alerting others that you were part of the world of the nerds. We loved it. Our slide rules were objects of devotion. We lubricated the slide scale with talcum powder. We adjusted the clamping screws to get the slide scale to move with just the right amount of friction to maximize speed while minimizing misalignment errors. And there was a reason they were so precious, beyond just bragging rights: Using a slide rule imbued a sense of the size of just about anything relative to anything else. My slide rule changed my life. Just by moving the slide against the body, I could quickly navigate the scale of physics, from atoms to the whole universe. It was all at my fingertips.

❧ ❧ ❧

One memorable day in high school, back in 1972, a kid from a technical family came in with a newfangled Hewlett-Packard 35, the very first pocket calculator. The "35" derived from it having 35 keys. It didn't just multiply and divide; it could also find sines and cosines and square roots. It could find the elusive "natural logarithms" of numbers. Wow. This was the precursor of all the other pocket calculators that followed. It was also the beginning of the personal electronics revolution that eventually led to home computers, laptops, and smartphones.

Now, we nerds took, and still do take, great pride in having a deep, personal feel for numbers. We have a sense of how big a number is supposed to be. By this I mean, when we are working numbers—even really big or small ones—we have an immediate sense in our heads where the decimal point should fall. In my day, we believed it was because of our use of slide rules. On a slide rule, 1.7 looks exactly the same as 17, 0.17, 170, or 1.7 million. So "slide rulists" like us had to keep careful track of the powers of 10 when figuring out how big the answer would be. We had to feel the numbers in our bones. With electronic calculators, you don't have to do that. At the time, letting a small box of circuits do the work felt a little bit like cheating to us. And get this: If you multiplied 9 times 9 on the HP 35, you got 81 (all good), but if you did 9^2, or 9 squared, you got 80.999999. Somewhere there was a tiny rounding error in the electronic logic. Ooooh, I thought . . . I'm keeping my slide rule.

The upshot is that my pals and I were not particularly impressed with modern calculators, at least not at first. We thought they were unnecessarily expensive and obviously fraught with shortcomings.

The original HP 35 cost $395, roughly $2,300 in today's dollars. Ouch. I took my trusty Pickett model N3-ES slide rule from high school to college. When I started engineering school, everyone still had a slide rule; most of us treasured them. My magnesium metal Pickett was said to be the same design that the *Apollo 11* astronauts took with them to the Moon, just in case they had to check some numbers along the way. I found out later that my model is actually a little more capable than the one NASA provided. It has a few more scales than the ones that went into space and back, meaning that it could perform a wider range of calculations. See? I could have been an astronaut!

When change finally came to the nerd world, it arrived fast and hard. By my recollection, it was the winter of 1975, when everyone went home for the holidays and everyone came back to school with an electronic calculator. Whether it was Christmas, Hanukkah, or just winter break, everyone had parents who saw what was coming and decided that holiday gift giving was the perfect opportunity to get their kids up to date. My first machine was a Texas Instruments SR-50. And do you know what "SR" stood for? "Slide rule." The manufacturer was practically shouting, "This thing is as good as a slide rule!"

If this is the point where you are waiting for me to wax nostalgic about the good old days, well, I'm about to disappoint you, dear reader. I can admit it now. The SR-50 was not just as good as a slide rule; it was better. It could do hyperbolic sines and cosines, for crying out loud! And there was something else important about the electronic calculator: It was more democratic than the slide rule. Making it easier for people to work with numbers and grasp scientific concepts was fundamentally a victory for knowledge and information, which in turn was fundamentally good for nerds. On the surface, sure, my buddies and I resented that it let more people into the nerd club without their having to learn the intricacies of the slide rule. Deep down, we understood that having more people in the nerd club was a good thing. We didn't see ourselves as outsiders; we thought we were the ones who had the best, most honest way of looking at the world (I still think so).

Pretty soon there were plenty of people who didn't know what a slide rule was for, even among the nerds. Human-built technology had delivered a more efficient way to crunch the numbers that define the results of scientific study that guide engineering solutions. By the way, I don't have my Pickett N3-ES anymore: It's in the Smithsonian Institution's collection. Archivists came and "collected me," and now my slide rule is preserved for future generations in some safe place somewhere. That's the perfect place for a device that helped get us where we are now but was not a final step. The slide rule naturally joins the astrolabe, sextant, and other tools that advanced the cause of science until science outgrew them.

Today it can be hard to comprehend how engineers ever did their jobs without computers and electronic calculators. I often joke about the black-and-white footage of early rockets falling over and exploding— that it happened because the only thing the rocket scientists had to work with back then were slide rules. But the truth is, they did do their jobs. Not just the NASA engineers; I'm talking about everyone in the pre-electronic era who found effective ways to expand the range of human comprehension by manipulating numbers in ways that the brain alone cannot do. I'm looking all the way back to the Sumerians who developed the abacus something like 4,500 years ago. Or William Oughtred, the Anglican cleric and nerd cult hero who invented the very first slide rule in 1622.

Throughout history, scientists and engineers (whether or not they called themselves those things) used the newest available technology to quantify their world. So despite my deep affection for my old devices, I feel more than fine about the near-extinction of the slide rule. It disappeared from the classrooms and the math clubs not because science doesn't matter to us anymore, but because it matters so much. We have vastly better, more powerful ways of manipulating numbers these days; it would be a little weird if we didn't use them.

Anyone in the world with an Internet connection can now get access to countless advanced math programs and easy-to-use software. You type in something like $y = \ln(78)$ and instantly out pops $y = 4.35671$ (and lots more digits, if you want 'em). Many of them are free, or at least a lot cheaper than my old Pickett N3-ES slide rule, and they can do things that would have blown the bow tie right off my youthful neck. Any time calculations become easier, scientific investigations become easier, too. The implications are both practical and cosmic. Medical studies will be increasingly reliable because the statistics will be more comprehensive, for instance. At the other end, astronomers may soon figure out the nature of dark matter because they will have access to zettabytes of data on the motions of stars.

The passion I tapped into back in high school was all about grasping toward those kinds of possibilities. The slide rule was merely a means to an end: an end in which a group of 10 numbers and a few dozen mathematical operations can let the human mind wander across space and time, to commune mathematically with everything in nature. These days that passion is more accessible than ever. Citizen science projects let anyone participate in climate research, scan distant galaxies, study the microbes in your gut, or listen for possible signals from alien civilizations. Free online courses teach advanced number theory. This is part of what I referred to earlier as the age of unprecedented access to information.

The slide rule provided a tangible bond for me with my fellow nerds, and that particular experience is sadly lost. Sure, I wax nostalgic about it from time to time. But I'm glad that there is one fewer barrier standing between us and a deep understanding of the world around us. My challenge to you now is not to fixate on the iconography of the slide rule (and all the other things like it that have become geek chic) but to look forward and, as Sam Cooke sang, "know what a slide rule is for." Play with numbers. Notice patterns. Apply what you know about the world in all sorts of places. Feel your place in the universe, and feel the universe that is inside you. Contribute to the scientific process if you can, or simply read and listen and revel in it. If more people engage in the deep comprehension that comes with mathematical thinking—what a wonderful world that will be.

CHAPTER 5

The First Earth Day and National Service

On April 22, 1970, I rode my bicycle to the very first Earth Day on the National Mall in Washington, DC. To anyone full of nostalgia for the disco era, I must remind you: Those were not really the good old days. In many ways Washington was a city divided, even more so than it is today. The 1968 riots following the death of Martin Luther King, Jr. left sizable parts of the city center physically and economically devastated; an unofficial but very visible boundary separated the relatively wealthy, white part of town from the relatively poor, minority-dominated one. The United States as a whole was mired in the behated and increasingly deadly war in Vietnam. Young people like me lived in fear of receiving a low draft number. Yet when I arrived for the Earth Day celebration, I felt a hopeful sense of unity.

The world was careening toward crisis, yes, but I was going to meet up with the people who were ready to do something about it. And

I was going to have a chance to be a part of the solution. I was so into it all that I fashioned a cardboard sign, held on my lower back or posterior with twine. The sign said, "Pedals Don't Pollute." The letter "o" in "Pollute" was the squeezed Greek letter theta, which had become the Earth Day symbol, like this: Pedals Don't Pⵁllute. Laugh all you want; what happened changed history. Nationwide, 20 million people showed up to help create the modern environmental movement. We are still living in the aftermath.

I locked my Schwinn Super Sport bicycle to a flagpole at the Washington Monument, just as you would if you were riding around in a small town. (If you tried that today, your bike would probably be taken to a remote location, x-rayed, and destroyed. Such is change.) Then I joined the thousands of others making their way to the National Mall. On the Mall side of the US Capitol Building was a huge stage; over the course of the day, a succession of speakers there described in chilling detail the ways that humans were harming the planet and urged us to reform our environmentally evil ways.

In those days, we were all intensely concerned about pollution. Just a year earlier, Cleveland's Cuyahoga River had essentially caught fire when a large oil slick on the water near the Republic Steel mill went up in flames. Soon, that river fire became an emblem of industry run amok. I remember riding my bike near the Potomac River around that same time and being incredulous that there were people out there on boats. It seemed impossible to me that anybody would voluntarily get near the Potomac, let alone on it: "Isn't that water too polluted to put valuable boats on or in? You can't be serious. What if a boat driver were to get some of that river spray in his mouth? Wouldn't he be dead in a few hours or even minutes?"

If you think today's environmentalists are doomsayers, you should have heard the Earth Day speakers. The message that I came away with was "Humans are bad." "Don't drive a car." Also, "Don't waste water—so wear dirty clothes" (like any good tree-hugging hippie). The overall lesson seemed to be that humans are bad for other living things, as well as themselves—us. Scientists were just starting to come to terms with the scope of our impact on the planet. The word "ecosystem" was relatively new, as was the field of ecology. It wasn't hard to see the overall trend, though. Living things interact in predictable ways, and we were seriously messing with those interactions. I might be exaggerating the mood a bit; the views of a teenager can do that. But to me, the warnings all seemed logical and conclusive. We couldn't keep on going the way we were going because we were destroying our world. Those views struck many as extreme, and they reinforced the idea in certain quarters that Earth Day was organized by that imagined bunch of dirty hippies.

I also got another, much stronger message that day: We have a collective responsibility and, along with it, the power of collective action. The majority of the 20 million people who showed up were ordinary people from all classes and cultures who were deeply concerned about the environment. When you added up that much concern, it was extremely influential. Factions in the US Congress quickly found common cause with President Richard Nixon and agreed to establish the Environmental Protection Agency (EPA). This is a piece of environmental history that is too often overlooked: The government agency most responsible for clamping down on polluters and making this country more green was born under a conservative Republican president. Equally impressive, the legislation creating the EPA was enacted just 8 months after Earth Day. Partisan divisions and gridlock on environmental issues are not necessarily a given.

Since its inception, the EPA has continually fought the good fight. Its mission is to protect human health and the environment. The agency wages a continual battle for the public good as industries and individuals seek to externalize their costs. "Externalize" is an economist's term that means "make someone else pay for it." External costs are the consequence of the basic truism that you can't get something for nothing. In general, polluting is easier than not polluting; otherwise, people wouldn't be doing it. As a result, there will almost always be some expense involved in being cleaner and more benign. Those costs run headlong into another truism: People don't like to pay for something if there's some way they can stick someone else with the bill. Industries and individuals fight one another all the time these days to externalize the costs of living. The problem is, *somebody* has to pay for the services and quality of environment that we all want.

In the case of the environment, we all pay for waste-treatment facilities that have to include systems to get rid of all the dirty things we produce, from industrial solvents and debris to our food preparation and digestive by-products, along with the dirty motor oil that your neighbor decided to pour down the storm drain. If the power company on the Potomac River discharged an effluent so hot that it killed the fish, the rest of us would not have to deal with rotting fish carcasses downstream.

Come to think of it, we did. As a Boy Scout, I once paddled a canoe through a slurry of fish corpses on the Potomac River. It was an eye- and nose-opening experience. The fish had died because, at the time, the local power company was not required to cool water from the generating plant before that overheated water was pumped back into the river. When the company eventually addressed the problem by installing expansive cooling systems, it charged us all higher electric fees to offset the cost of the cooling equipment and the extra real estate it was built upon. In this example, the company redirected its externalized costs to the people who pay electric bills, rather than onto the fish and the people who would otherwise have had to deal with disposing of the rotting fish carcasses.

Economists call situations like this the "tragedy of the commons," a reference to the shared grazing land in the British Isles. Some people might decide to graze one or two more cattle than is allowed in the commons agreement. A little cheating doesn't seem like it would cause any harm. But if most or all the people do that, pretty soon the commons will not be able to support everyone's grazing animals. In other words, common resources will vanish unless people share responsibility for tending to them. So how do you ensure that the commons, and any other shared resource, isn't being exploited by someone or some agency that puts its needs ahead of the common good?

This is where the unfairly vilified term "regulation" comes in. As an engineer, I think of regulations as being like a complicated modern factory, complete with robotic welders, conveyor belts, sorting devices, and so on. You could easily get carried away when designing a factory like this, buying too many machines and giving the assembly line too many twists and turns. That would waste money and reduce efficiency, and perhaps create serious snags in the production process. On the other hand, there is a critical minimum set of components that a factory needs to operate. You can't arbitrarily step in and decide "I don't like the look of that robot, so let's not bother doing that part of the welding process." And once you've worked out an agreeable set of components to keep things moving smoothly, you have to remain alert to malfunctions. You cannot let the belts wear out; if you don't maintain them, they will break and bring the whole factory to a halt.

So it is with environmental rules. We want all the ones we need, but no more. Too many and we start to interfere with innovation and economic growth while providing little practical benefit. Too few legal protections pose a more critical risk, however. If we don't pay enough attention to our shared responsibilities, we can allow serious harm to humans and to the other species we rely on. Environmental pollution, especially the carbon dioxide emissions associated with climate change, is the tragedy of the commons writ large—as large as can be. Regulation is really just a legal analog of smart engineering. It formalizes our ideas about how best to take care of the planet (and ourselves) so that the whole system keeps running smoothly.

It is fashionable in certain circles these days to talk about government agencies as if they are autonomous devices that merrily go about pursuing their own agendas. That is about as absurd as criticizing a factory foreman for caring more about the machines than for the final product. Taking care of the machines is inseparable from taking care of the production process. Similarly, agencies like the EPA were created to take care of the planet and avert tragedy. I can tell you, as someone who was on the scene at the first Earth Day, there could hardly be anything more human than the worried, passionate, diverse crowd that gathered there to agitate for a cleaner planet. The EPA, and all the other agencies that do like-minded work, is composed of true ardent nerds working hard to focus on the big picture and to serve the needs of the many rather than those of the few. When those agencies are under assault, we are *all* under assault, and we need to defend our common interests.

The EPA is the American people, and we are the EPA. At least, we should be.

$\diamond$ $\diamond$ $\diamond$

It may be hard to recall, but for a long time the idea that humans could change the entire planet seemed ludicrous. The prevailing attitude since the beginning of the Industrial Revolution was that Earth is so huge, and humans so puny, that the most we could do was local damage.

That view started to change in earnest only in the 1960s, for several reasons. Astronomers began to compare the climates of other planets with the climate of Earth. When the first photo of Earth, looking delicate and alone outside the window of the *Apollo 8* command module, was sent around the world late in December 1968, the impact was huge. Those first personal images of our planet from afar caused a dramatic change in perspective.

And then there was that first Earth Day. It wasn't "Local Cleanup Day" or "National Environment Day." The rally was about getting us all to think of our planet as a single enormous ecosystem—the Earth, the global commons. There's an inherent morality in this everything-all-at-once attitude. We are all responsible for the global commons. We all have to look out for our neighbors, and now it's clear that those "neighbors" might be halfway around the planet. Whether you are a scientist or an artist, a very important leader or an ordinary citizen, you have an obligation to pitch in for the greater good. Soon that way of thinking seemed not just reasonable but also obvious . . . well, to most people. So yes, Earth Day was effective. It had and continues to have a considerable impact.

After its initial rousing success, Earth Day has been celebrated on April 22 every year. I rode down to a few more of the events when I was in high school in the early 1970s. Like so many things in our society, Earth Day gatherings became more organized and more commercialized. From one perspective, it's easy to wonder what good my short bike rides to the gatherings really did. One could argue that my attendance at those rallies made no material difference to the politicians calling the shots in the Capitol Building and the White House. I'd dispute that view, though. I think the continued crowds, year after year, helped maintain support for the EPA and the many other, less visible state agencies that magnify its work.

I can tell you for certain that Earth Day motivated me. I was convinced that we were headed for trouble as a species unless we could start using our brains more rationally, and it shaped how I approached my own environmental impact and goals for the future. Ever since that first event, I have done whatever I can to fight the good fight and to get you to fight alongside me. Support the causes you believe in. Show up at rallies. Find your community, or help create one if you can. Share support and inspiration for nerdy compassion and responsibility. Stand up and be counted as an active, rather than a passive, member of your democracy. And use your time at these gatherings to find out what courses of action you can take next that can continue to make a difference at the local level and beyond.

A lot of what I have done over the past 4 decades—including writing the book that you're reading right now—was inspired by the need to help people understand what it means to be a global species. We humans know we can change the world because we're doing it right now—but so far we're doing it mostly by accident. The challenge is taking responsibility for our actions, assuming deliberate control of the change. We're all in this together. There's nobody else who can pick up the external costs. Where, exactly, would you send the bill? I don't think any of us has that address—ha, ha, ha (?).

Nerds don't give up when there's a problem to be solved, and so I've kept working on my climate message, redoubling my efforts to try everything all at once. I describe climate change in my kids' books and made a point to do climate change demonstrations on the *Bill Nye the Science Guy* show. I've spoken at Earth Day events in Washington, DC, at the invitation of both Democratic and Republican presidents. I managed to host a special on the National Geographic Channel with Arnold Schwarzenegger about environmental destruction and global warming.

One thing led to another, and President Barack Obama had me as his guest at the 2015 Earth Day celebration. We went to Florida to call attention to environmental issues there and to celebrate a redesigned and rebuilt system of public works, bridges, weirs, and roads that together are helping manage a major redistribution of surface water in and around Everglades National Park. The Everglades has an exotic, delicate ecosystem full of species that are found nowhere else on Earth. If you want clean water in South Florida, you want that surface flow to be filtered by the complex chemistry of the living systems in North and Central Florida. All this is worth preserving, and the efforts currently underway are impressive. That is part of what President Obama wanted to discuss with me.

The Everglades restoration is a metaphorical drop in the bucket, however. As global temperatures rise, sea level is rising, too. Ocean saltwater may soon inundate large parts of Florida, including the Everglades. Humankind, the United States especially, has done very little to address climate change. We need to press on and apply nerd thinking much more broadly. I try to do my part on the individual level. I recycle. I bike on local errands and to local business meetings. I drive an electric vehicle. I have solar panels on my house. I have a solar hot-water system. Most of the people I know take all kinds of personal actions to help out, too. But that's not enough. We need to put on our Big Data goggles and remember the power of applied science and engineering, especially when we harness it together.

I ask myself—and I hope you will ask yourself, too—what will turn the tide this time? What will motivate the United States and the rest of the world to settle in and get to work addressing the effects of our warming world? I am quite sure that it will take all of us working collectively toward a common cause. Crazy, large-scale idealism isn't crazy at all. It really can be put into practice. Air and water are much cleaner than they were in 1970 across most of the developed world. Rivers no longer catch fire in the United States. Washington, DC, is a much wealthier and fancier town than it was back then; the neighborhoods burned out during the riots are once again vibrant communities. But the kind of urgency that led to the rapid adoption of environmental regulations is lacking this time around. The crisis mentality of 1970 wasn't pleasant, but it got results.

The responsibility lies with every one of us. Whether you frame the problem in terms of the tragedy of the commons or in terms of everything all at once, the message is the same. We have to address climate change every way we can, using the very best scientific methods. We have to reduce our waste and do much more with much less. We have to develop clean-energy technologies. We have to provide access to these emerging technologies to everyone we can. And we have to do it not just one by one, but also as a nation and as a planet. It's a moral imperative, and it's also imperative for our survival.

Rekindling the sense of responsibility that so many of us felt at the first Earth Day is one of the great challenges we face in putting today's environmental ideas into action. To a large extent, we are victims of our own success. The environment is much cleaner and the economy is much stronger across much of the Western world. Reports of heat waves and flooding do not stir a sense of immediate outrage the way a flaming river did. With the growing cynicism toward government institutions, people are not as invested in public service as they once were. There's no one solution to all of that, but there is one idea that I keep thinking about . . .

Perhaps because both my parents were veterans, I took an interest in joining the US Air Force when I was in college. I considered myself a patriot, I was fascinated by jets, and I wanted to do something for the country—all the more so if it helped pay for my college education. So in 1975, I took the first steps toward becoming part of the Reserve Officers' Training Corps (ROTC). While I was waiting to take my physical at the Air Force base in Rome, New York, I got to talking with a few Air Force pilots there. I asked them how often they flew, and they said, "Oh, every couple months, maybe every six weeks." Then I asked what they did between flights. They replied, "We memorize procedures, do paperwork." The pilots and their skills were languishing.

What I heard from those nonflying flyboys sounded like a drag, so I didn't stick with it. I delisted without consequence. But I often think about how different my life would be if I had stayed—if there had been an exciting role waiting for me, whether or not it involved flying fighter jets. I might have ended up in a lifelong career of service to my native country. At the very least, I would have had formative experience in public service. I would have been keyed in to issues of national and foreign policy in a completely different way than I am now. Perhaps I would have found my way to the ideas of collective responsibility and collective problem-solving more quickly.

Then I think about how it was for my parents. During World War II, my father was captured and held in a POW camp. Once you're in a prison camp, you pretty much have to stick with it; there's no delisting. Meanwhile, my mother worked for the Navy, solving puzzles to help win the war while wondering about her boyfriend who had disappeared from a remote Pacific island. For them, public service was not optional. They had no choice but to work side by side with people from all walks of life, everyone pursuing the same goals. Blue-collar workers fought alongside guys in law school. Women from the poor and rich parts of towns riveted airplane wings working elbow-to-elbow. Everyone pitched in, because the situation was so serious, and there was no one who did not perceive what has become a popular term: the "existential threat." The very existence of everyone's homeland was at stake.

Could we recapture some of that spirit with a new national service in the United States? I'm picturing a system in which every US citizen has to serve for 1 year before he or she turns 26 years old. It would be international in scope, so you could serve either domestically or abroad—somewhat like a global version of the Works Progress Administration (WPA) created during the New Deal years. You could join the military, but you could also serve on a crew erecting wind turbines and transmission line towers. You could help develop low-cost photovoltaic cells. You could assist teachers in schools. You could tend to the elderly. Or you could go overseas and work to provide clean water, renewable electricity, and Internet access to citizens of developing nations. You could be an ambassador of Western culture and nerdy knowledge, building collaboration and trust across international lines. Certainly you wouldn't feel as though you were wasting time on make-work memorization.

What I'm imagining is something bigger than the Peace Corps or even AmeriCorps. It would be a full year, not just a summer, and it would be mandatory. It would be law, and everyone single one of us would have to serve. You could serve right out of high school, after college, or between jobs. But if you were planning to continue on to graduate school, then somewhere during your academic career you'd have to serve. I imagine many grad students would choose the time after a master's degree and before a PhD. The collaborative spirit of the first Earth Day would be part of the law of the land, regardless of your academic path.

Some of you may bristle at this idea, but I believe that a program of national service would change us for the better and break down a lot of the partisan tropes that divide left and right. Liberals are broadly suspicious of government coercion but are very supportive of the idea of humanitarian programs. Conservatives are broadly suspicious of government social programs but are very supportive of the idea of service to the country. Across the board, there is deep distrust of federal institutions right now. Okay, so suppose each state enacted a state service program. Participants might get tuition to the best state schools in exchange for their time. Then if Indiana, say, had a state service, it would put competitive pressure on Ohio to enact the same sort of legislation. State by state, local services would become mandatory and ubiquitous. Such programs could begin a healing process. I'm hoping to plant a seed here so that a leader might nurture the idea and someday bring us together to serve.

Now look, I'll remind you right here: I did not serve a single day in the Peace Corps or in the military. But as I compare my experience with that of my parents and their friends and contemporaries, I feel that kind of service could have done a lot for me and would do a great deal for others, too. It could help bring us together, engage us with the rest of the world, reduce our fear of outsiders, and embrace the future. A national service program would cost (tax) money, but the funds spent would immediately find their way back into the economy. The millions of people enrolled in the service would rebuild infrastructure, move us to renewable energy, and work with citizens of other countries to expand access to clean water and digital information. Earth Day was initially considered part of the counterculture. This could instantly be a part of the mainstream.

Imagine along with me here. Someday soon, we could see young Americans hard at work building wind turbines, installing photovoltaic panels, and setting up water desalination plants where they're needed. They could be expanding renewable energy in Appalachia or rolling out cellular Internet access in Ethiopia. Every step along the way, they would be chipping away at the partisan and cultural barriers that we've erected in so much of American society these days. Each service program would contribute a fraction to fight climate change but could contribute a great deal to a sense of shared purpose—before we get there the hard way, by plunging into a full-on crisis.

Whether or not my "national service corps" concept ever gains any traction, the important point is that there are many ways to recapture the spirit of 1970 and rise to the challenges of today. If Nixon could create the EPA practically overnight in the aftermath of Earth Day, then the current president and the current Congress can certainly make great things happen, too. We just have to agree that it needs to get done and then get going, full speed into the future.

CHAPTER 6

How My Parents Quit Smoking

I hereby claim that if you want to get grown-ups to stop merely talking about important issues like climate change and to start doing something about them, one of the most effective things you can do is get their kids involved in the process. Young people are curious, idealistic, and eager, making them the perfect instigators of science-based change. They also have a huge tactical advantage: They are living at home, so they can perform their persuasive feats on their parents from the closest possible range. I became convinced of these things because of a remarkable adventure I had when I was 12. And it was all to do with cigarettes.

In the 1960s, when I was growing up in Washington, DC, a great many of the adults I knew smoked. In fact, it seemed like almost every grown-up smoked back then. My mom and dad smoked. Even my neighbor's dad, a medical doctor, smoked. And like virtually every smoker

I've ever met, they all talked about wanting to quit. They enjoyed smoking after a fashion, but really, deep down, they wanted to stop. Since the 1950s, there have been individual and class-action lawsuits going on against tobacco company executives. I remember them being featured on the television news shows I watched with my family. The execs kept claiming with earnest expressions that they never heard of any connection between smoking and cancer.

My parents absolutely did not want me to start smoking, and they remarked often about how hard it was to quit. Planning to quit smoking was as much a part of the culture as the smoking itself. It was routine to hear radio ads featuring the voice of a firefighter and sound effects of a crackling fire and sirens. The firefighter would intone ominously that this needless destruction was caused by "someone smoking in bed." But the one message that really made an impression on me featured William Talman, the actor who played District Attorney Hamilton Burger on the hugely popular television drama *Perry Mason*.

If you're not hip to this old reference, note how many recent and ongoing television shows feature lawyers: *L.A. Law*, *Law & Order*, *Boston Legal*, *The Practice*, *Ally McBeal*, *Better Call Saul*, and *The Good Wife* (by no means a complete list). Well, *Perry Mason* was the first of the genre. Note that the name "Hamilton Burger" sounds a lot like "hamburger," because this character got ground up every week by our hero Perry Mason, the uncannily brilliant lawyer. Perry (as we call him) always outsmarted the DA and the cops, and he extracted the confession of the real murderer just as the show was going to its last commercial break. Perry brought truth and justice through his relentless application of logic and human insight. I loved this show—*loved* it.

So when Hamilton Burger—I mean William Talman—came on my television and warned, "By the time you see this, I will be dead. Please don't smoke!" I was riveted. I felt driven anew. I had to do something to save my parents. I'll admit, there was also a twinge of "These grown-ups with their 'do as I say, not as I do' behavior—it's not right." But it was, I suppose, a larval version of my "change the world" impulse. I was just 12, so I had the more modest goal of changing our household. I harassed, er, reminded my parents repeatedly about how they had both promised to give up smoking. And their old soundtrack would repeat: They would respond that they really wanted to quit, but it was really hard, and then usually urge me once again to never, ever start smoking.

As a matter of parental course, they wanted to "enrich my summer"—a great bit of parentspeak for "You're too old to send to sing-along camp and too young to get a bona fide paying job, but it's time to get out of the house, Bill." So for my *enrichment*, my parents arranged for me to attend a summer program at the Smithsonian Institution in downtown Washington, DC, an easy commute from home. I got to learn all about oceanography, which at the time seemed like the coolest thing possible. For a budding science guy, the summer was a success. For my nicotine-addicted parents, they assumed that they'd bought themselves a few weeks of relief from my meddling, do-gooding ways. Oh, how wrong they were.

To get to the Smithsonian, I took the bus downtown and got to know a group of guys, mostly older, who were in the same program but were a couple of years ahead of me in school. We got to talking and became friends. At some point, we started walking to a slightly more distant bus stop because the route took us by a place of business on Pennsylvania Avenue called Al's Magic Shop. It was a well-known spot, legendary in the magic world, that ultimately stayed in business for 58 years. Along with your cut-off-your-finger effects, your easily com-pacted silk flowers, and all sorts of card tricks, Al Cohen sold what were called "cigarette loads." These small pyrotechnic devices look like the broken-off tip of a toothpick. You poke them into the end of a cigarette. A few minutes or even seconds after the unsuspecting smoker lights up (depending on how deeply the load is embedded), the cigarette explodes—dramatically. Afterward, the cigarette paper is rolled back like a perfectly peeled banana. What's left of the cigarette looks like an exploded cigar in a Warner Bros. cartoon, the kind that turns Daffy Duck's face black and knocks his bill sideways. The whole process is destructive and shocking and just plain wonderful. The cigarette loads are basically perfect tools for adolescent mischief.

So as a creative young man prone to experimentation, what do you think I did? I bought some cigarette loads and carefully loaded them into my parents' cigarettes. That would teach them to keep smoking after they had repeatedly promised me that they would stop. I was an aspiring engineer, and I tinkered a little. I did not exactly follow the instructions, in which the manufacturer recommended emptying some of the tobacco and putting the load right at the end of the cigarette. Instead, I used a straight pin from my mother's sewing supplies to ever so carefully push the load deeper into the cigarette so that you could not see the load; it was invisible. There may also have been some physics implications that I had only roughly considered when formulating my plan. With the cigarette load fully embedded in the tobacco and constrained by the "hoop stress" of the paper, perhaps its explosion is ever so slightly more forcefully contained.

The whole caper was surprisingly easy to pull off. Cigarette loads are very small, thin as a grain of rice, so concealing them was simple enough. My dad took out his cigarettes when he emptied his pockets each evening. He left them out in the open on his dresser, alongside his wallet and keys. There was no trouble getting my hands on his stash for a few private minutes.

One night my parents went to the next-door neighbors' house for dinner. That was not an especially uncommon occurrence. Our two families had become good friends. It was summer in Washington, DC, and as you've probably gathered by now, Washington summers tend to be stupidly, oppressively hot and humid. Those lazy, late-sunset evenings were fine times for laid-back socializing. Since we did not have air-conditioning, every window was wide open to let in whatever almost-cool breeze we could create with electric fans, fans, and more fans. The same was true next door. So when the cigarettes exploded in the house next door, I could hear every single bright, deeply satisfying *pop*. Woo hoo!

The grown-ups were surprisingly good-humored about it. They could pretty easily guess who was behind the prank. I could hear them laughing but pressing on, trying cigarette after cigarette over their after-dinner coffee and dessert. *Pop-pop-pop*. Every few minutes there was a new attempt to light up, and a new surprise. As it turns out, I had incorporated a fabulous extra detail in my explosive experiment that was unexpectedly effective. The time from the lighting to the detonation of each cigarette was not uniform; it was randomized, depending on how deeply I had pushed in each load. Once the loads started going off, neither my parents nor my neighbors could tell which cigarette would pop off next, or what would happen when they lit up another one. Since I was working on a budget, a few of the cigs were unloaded altogether. The cigarette–smoker interaction was unpredictable, which made it especially psychologically disruptive.

If you want to train your lab rat to tap a lever with his or her paw (there are rat males and rat females—you can look it up), you rig up a mechanism to give the rat a sunflower seed or similar delicious reward each time he or she steps on the lever. If you want your rat to avoid a metal plate, you give your rat a mild shock whenever she or he sets a paw in that area. That's the kind of stuff that Ivan Pavlov figured out. But if you want to make the rat go a little nuts, you give the reward or jolt only now and then, and do it at random. Remove certainty and replace it with nerve-jangling chaos. That is what I, the budding and not-quite-evil genius, had done to the grown-ups. Bwah, ha, ha, ha, haaaa . . .

Despite their encouraging laughs, my parents were not entirely amused by what I had done. They had a talk with me about my explosive stunt, about how it startled their friends, about how it was wrong to mess with their property. But as for my motivation, what were they going to say? "We want to quit—but don't make it so we are afraid to smoke."

"Why not, Mom? Why not, Dad?"

"Well, um . . ."

As a science-minded guy who has managed to live in our society for 6 decades, I am deeply aware that addiction is serious business. Burning tobacco, especially the nicotine it releases, affects smokers' brains. Addicts of all stripes start using the drug or engaging in the addictive activity to feel good, but eventually they have to keep going to, as the saying goes, "not feel bad." The effects on the brain alone are amazing and troubling. But along with the chemical aspect of addiction, there is also the behavioral part. Smoking becomes a habit, almost a ritual. The same thing can happen with alcohol, other drugs, even unhealthy eating or gambling.

For example, my parents and our neighbors routinely lit up their cigarettes after dessert. It's what they, and a lot of people like them, did back then as a normal part of being social. It's what we saw in the movies and on television (even, yes, on *Perry Mason*). But the effect of those cigarette loads was powerful and lasting, even beyond what I had hoped for. My prank did more than startle my parents. It shook them up. It made them acutely aware of their addiction, to the point that they managed to quit.

With this in mind, it is reasonable to conclude that by breaking a habitual behavior, we can start breaking free of the addiction. I wasn't plotting, at least not consciously, to trigger some kind of Pavlovian response in my mom and dad. I merely wanted to remind them, in an emphatic way, how nasty cigarettes are. Okay, I also liked the idea of making them jump, and apparently they did that quite expressively. (They were next door and I couldn't see their reactions in the moment.)

The night of the exploding cigarettes led to my parents quitting for real. My cigarette loads apparently accomplished what my years of youthful nagging could not. They could never be sure after that just how many cigarette loads I'd invested in, let alone where I had gotten my supply. And what if I decided to strike again, despite the talk? They had no way to know whether the next cancer stick they were about to light might turn into a dramatic and disruptive event. My parents were shaken out of their routine in a way that interfered with their social addiction.

I realize that my prank was not the whole reason they quit smoking; it probably wasn't even the main reason. But it did help them change their behavior, and that was enough to nudge them into giving up a habit that had stuck with them for many years.

I should add here, loud and clear, that I do not generally advocate using explosives on your friends and family, no matter how virtuous your motives. I don't want to lose any potential readers, plus it's just so dreadfully . . . impolite. I loaded up my parents' cigarettes only because I was quite certain those little charges from Al's Magic Shop were harmless. What I most definitely *do* advocate is being an activist for change in your own home. When you are trying to change the world, it is often most effective to start with the people you personally influence the most. And if you can create that change using a scientific process, all the better.

My detonating cigarettes were part of a much broader movement, as it turned out. William Talman died a few months later, at the too-young age of 53, but the antismoking movement continued and grew. Smoking is far, far less common these days, at least in Western societies.

Social norms are changing. It's banned in bars and clubs in much of the country. In New York City, many restaurateurs made grim predictions that business would plummet if patrons were not permitted to sit around and smoke with their coffee after dessert. That's not what happened, though. Instead, more people stuck around for longer and ordered more food. Restaurant revenues went up. Today, smoking rarely even shows up as a cool thing in modern movies, unless they are oh-so-hip swing-era period pieces.

Just as in the Nye household, a lot of those changes began on a local and personal scale. Family members urged their loved ones to abandon a dangerous habit. A relatively small number of very engaged people started making the social environment different, and eventually a large number of people started behaving in new and much healthier ways. This is a really big deal. Millions of people stopped smoking, millions more never started, and as a result, millions who would have died of lung cancer or suffered from chronic obstructive pulmonary disorder will not. That's a lovely, loud echo of the little noise that adolescent Bill managed with a few well-targeted cigarette loads.

All the more reason to get our kids exposed to science and primed to carry out their own explosive acts of progress. My parents were a bit angry with me about the cigarettes, but they understood what I was up to, probably even more than I understood it myself. By the age of 12, I had picked up enough about the health hazards of smoking that I knew what was at stake. I wanted my parents to live. That I got joy out of blowing stuff up was one thing, but that I did it out of love was another, bigger thing. I see the same scenario happening all the time now: children nudging their parents to recycle, to stop using plastic bags, to cut down on waste, and to learn about making the world a better place for the future. They are working with a potent mix of idealism and personal concern that looks very familiar to me.

High-quality science education is therefore a doubly powerful formula for change. In the fairly obvious ways, it gives the next generation the information and the critical-thinking tools that are essential to rational decision-making; it inspires data-driven leaps of youthful imagination, whether or not the kids end up in overtly scientific careers. But in the less obvious style of my explosive cigarette escapade, it also empowers them to act as behind-the-lines agents of change.

By their very nature, kids are more open to new ways of doing things than their parents are. They are energetic and, for better or worse, they can be relentless. With some early-stage everything-all-at-once training, we can push more of that energy to the "better" side of things. I'm not talking about whining and screaming. I'm talking about the kind of persuasion that is much more effective when it comes from inside the home, just as happened to me with my parents.

Don't underestimate the power of family bonds. Every generation wants to make a better world for their kids, but every generation also needs kids to keep them honest. If we want to change the behavior of people, convince their kids why it's so important for their parents to do the right thing. With our kids, I believe we can get humankind to a better place. If young people embrace evolution and connect the fundamental idea in life science with the mutation of pathogenic bacteria and viruses, vaccinations will be universally accepted; public health will improve. If we can get young people to understand and engage in affecting climate change, older generations will likely do a much better job preserving Earth and its ecosystems.

Of course, young people are not the only ones who can evaluate evidence and work for a better world. We should all try to hold on to the energy and enthusiasm of youth. At all ages, we should aspire to our own little explosive acts of progress and constructive defiance. Let's go!

CHAPTER 7

Ned and the "THANKS" Sign

Although he loved driving, my dad, Ned, was not an especially good driver. To be honest, he was at times an actively bad driver. He would pull into traffic too abruptly and burden the guy or gal behind him to slow down to avoid a collision. Or he would veer into another driver's lane and remark, "He can see me," trusting that the other driver would take action. All good and charmingly old-fashioned, I suppose—unless the other driver *didn't* see him. Nowadays, the other driver seems to see you less and less, due to texting and apps and all the other distractions on our ubiquitous mobile phones. (This may change with smartphone-summoned drivers and the driverless vehicles of the near future.)

I should be thankful that those things didn't exist when I was a kid. I can only imagine how unfocused my dad would have been. Don't get me wrong; he wasn't dreaming up ways to play games or chat with faraway friends. He was continually checking his compass and his altimeter—yes, an *altimeter* that would report his height above sea level (a bit of info that amazed me as a kid). He was also something of an amateur inventor. Combining his fascination with gadgets, his proclivity toward politeness, and his frequent inattention to events on the road, my dad spent a lot of time thinking about ways to communicate with other drivers. Like a true nerd, my dad was concerned with ways to add to the greater good. Decades before anyone used the term "road rage," he recognized the dangerous consequences of driving while angry. And in his inimitable do-it-yourself style, he decided to do something about it. He was DIY before DIY was a thing.

My father's solution was the automotive THANKS sign: a message board to let himself be heard by other drivers when he could not actually be heard. It was more than a concept. My brother and I rigged it up, following our father's specifications. My father was fully capable of measuring the sign; fitting the die-cut letters carefully with a ruler, onionskin paper, and glue; and so on, but he had other things in mind. Making the sign became a project to occupy his sons in a family activity that spoke not only to the technical challenges but also to the greater lessons of courtesy, cooperation, and over-the-road efficiency. Earlier I mentioned family bonds. My dad was very clued in to that, or those. So my brother and I, according to his instructions, attached a hinged piece of Masonite board to the rear deck of our dad's Renault 16; mounted a screw eye in the ceiling headliner; and ran a length of fishing line from the sign to the rearview mirror up front. I even attached a little rawhide strip at the end of the line to give the driver a better grip.

Here's how it worked. Let's say my dad had changed lanes without signaling, or pulled right in front of another vehicle while leaving just enough space for the other driver to avoid him. To express his appreciation for the other driver's defensive reactions, he would pull a cord attached to the rearview mirror and a sign with 6-inch-tall letters would flip up on his car's rear deck. The word displayed was a simple "THANKS." Then the driver behind him, instead of being furious with my father's irresponsible behavior, might feel great about himself or herself: "Wow, how thoughtful of that driver I almost rear-ended! He is thanking me. And my word, I must certainly be a courteous driver myself for having let this guy whip in front of me."

On second thought, it's possible my father wasn't really all that bad a driver. My brain may be selectively holding onto the memories of times when I felt like our lives were hanging in the balance, or when a sudden stop smashed my 4-year-old nose against the steering wheel. But the thing I know for sure is that Ned Nye had a deep idea that if we all cooperated rather than competed for space on our nation's streets, traffic flow would be a little better for everyone. Today, you have a smidge of control over traffic by planning your trip through Google Maps, Apple Maps, or Waze (or whatever new software emerges between my writing this and your reading it), which take congestion into consideration when recommending a route. Tomorrow, your car may have an autopilot that communicates with other cars to adjust speed or switch roads to optimize traffic flow. Long ago, my dad had the idea of achieving the same goal not through technology alone, but with a lovely combination of human compassion and technology—namely, a sign on a string.

The THANKS sign worked beautifully. People behind us really did acknowledge my father's cleverness and good intentions, if not his sons' craftsmanship. After my father unexpectedly slammed on the brakes in front of someone, he'd flip the sign, and most of the time, the other driver would give a return *beep-beep* or a flash-flash of his or her headlights. When we received some acknowledgement of our good manners after each THANKS-sign deployment, a happy conversation broke out in our car. "Wasn't that great? He got it. He tooted his horn at us." (Aren't we clever, etc.) Maybe that was part of my dad's master plan all along. Either way, the sign worked well enough that I'm still admiring it in a circumspect kinda way. I like to think that it really was a meaningful step in the long march to optimize traffic flow.

A few years later, my brother went off to college. I worked alone rigging up a similar THANKS system in the two of my father's cars that came after that Renault. By modern standards, the engineering process was crude and cumbersome. Just the idea of drilling holes in your car is way out of most of our experiences. If signs such as this were to be created today, they'd probably be electronic, perhaps using a nice bright LED display. They would be wired or Bluetooth-connected, no fishing line needed. Such a thing could be rigged up to work like a news reader's or stage lecturer's teleprompter. With a system like this, the words would be configured in mirror text and projected from the car's horizontal rear deck up onto the sloping rear window, which would be half-metalized to display the words for other drivers to read while allowing a clear view out from inside the car. It would loosely resemble a commercial head-up display, but reversed to aim outward at other drivers. Wait a moment—I have to head out to my workshop . . .

So what if time has completely passed my dad's technology? The notion that you could improve the world by providing a readable thank-you message everywhere you went still strikes me as pretty cool. I admired my dad for thinking in terms of solutions to real-world problems. I admired him, too, for not just dreaming the idea but for actually building it and using it. Better yet, for getting *us* to make it.

With Ned Nye, those kinds of ideas just kept coming. He had another obviously brilliant (?) invention to improve auto safety: the pedestrian horn, a gentler alternative to the regular heart-attack-loud auto horn. Over the years, we owned two separate French-built Renault 16 automobiles. (Ned repected the innovation found in foreign cars at the time. Though having been a prisoner of war, he was not much for Japanese cars of any kind.) My father had me mount a bicycle horn, which in the invention's first manifestation was battery powered, under the hood of the car. The horn was activated using what would have been the handlebar button if the horn were mounted on a bike. My father hung the button from the choke knob (intake airflow choke? older reference; it's technical—don't worry about it).

When he was concerned that a pedestrian was crossing in front of him without looking, my father could press the button and alert the person. The resulting toot was loud enough to draw attention but soft enough that other drivers would either not notice or realize that the sound was not intended for their ears. Like the THANKS sign, the pedestrian horn was an attempt to do a lot with a little. Using what I learned in physics class, I rigged up a voltage divider circuit to power it. Batteries in those little horns go dead fast. It really was fun.

These inventions were my dad's way of pushing back against our more negative impulses, encouraging all of us to cut one another a little slack; he demonstrated this to the people he communicated with, and even more so with his family by involving us in the creative process. By doing so, we make the world slightly more efficient and distinctly less stressful.

Such a system is needed more today than ever, given the proclivity of modern pedestrians to cross the street while looking straight down at their mobile phones. In Pasadena, California, the city has placed durable stickers on the pavement right by the curb in crosswalks. The stickers say, "Look Up." I personally believe they should be interpreted to scream: "LOOK UP! Now!!" My dad may have been a calmer man than I. He was a homespun innovator doing what he could to make people a little calmer, a little more efficient, and a little nicer. The sign and the horn were chipping away at the discourteousy [*sic*] he saw spreading around him. Ned's solutions seemed odd at the time and, I'll admit, they still seem a bit eccentric today. No investors ever came clamoring to mass-produce either of them. In their own ways, though, they did change the world, at least for me. As far as I'm concerned, nearly 5 decades after Ned was thinking about ways to make the world safer for pedestrians, the rest of us are still catching up. Some modern cars come with both a loud horn and a soft horn.

What I got out of these experiences, constructing and using my dad's oddball inventions, was an appreciation for his strong belief in respecting others. It is almost instinctive to feel like you have a right to be belligerent when another driver cuts in front of you, compelling you to stomp on the brakes. It is all too easy to cut loose with unprintable obscenities when a pedestrian steps off the curb a little later than he or she should have, forcing you to wait to make your right (or in some countries, left) turn, or to stomp on the brakes again to avoid a collision. It's tempting to jam your horn aggressively when a driver in front of you put on his or her left turn signal a few moments after the light turns green. No point in denying it: You have transgressed in these ways, countless times. I'll confess right here, I have, too. And yes, I realize that you, like every single one of us, is a self-evaluated above-average driver. Nevertheless, I strongly suspect you have screwed up from time to time.

The current trend in automotive engineering is to reduce traffic accidents by taking control away from the driver. That's one way of eliminating the unruly human element, and I expect it will yield huge benefits—saving lives and freeing up personal time. I'll have a lot more to say about that a little later on. (I hope you keep reading.) My father had a more modest goal. He wanted to make a better human driver, a kinder and gentler person who is more in control of the emotional and safety experience of being on the road.

Wait—that's not a more modest goal; it's an even grander one. I mean, it's huge. It's a very everything-all-at-once kind of goal, in fact. Acknowledging and getting past the minor mistakes we all make from time to time, whether we are driving or walking, encourages us to focus more on the big picture of collective action and collective responsibility. The automobile as we know it could not exist without a whole system of rules and standards that we all agree on. Some of them are so ubiquitous that we rarely even think about them: stop signs, traffic signals, lane markings, and so on. Some of them are more explicit, like safety standards or the EPA emissions rules that keep our urban air breathable.

In a way, my dad was getting at some of the most fundamental questions about technology: What is it for? Whom should it benefit? What responsibilities do we take on by creating it? The THANKS sign was one small expression of his answer. Technology should improve the way we live—make it happier, safer, calmer, more productive—and it should benefit anyone who is exposed to it. Which is to say, in principle, everyone. Those may seem like obvious ideas, but look around and see how often they are put into practice. Many people and corporations create things without seriously considering their impact, or they willingly accept that some innovations will benefit only a lucky few and might even have a negative impact on the rest. They fail what I will humbly call "The Ned Nye Test."

Even if we decide to put computers in charge of every car and truck, and even if the whole system works flawlessly without the need to flash a single sign, my dad's sense of collective responsibility will remain just as important. We'll still have to maintain the roads and bridges. People will keep changing and moving, and we'll have to build new infrastructure and develop new types of transit to match up with those changes. Doing all that requires taxes, an effective government, and an involved public. As the saying goes, "the most important government position is 'citizen.'" To many of us, these aren't glamorous things—and let's be honest, neither was my dad's floppy sign—but they are essential. They are part of an unspoken pact we all make to be kind to each other, to accept some responsibility in exchange for promoting the greater good.

When we keep that pact, we make progress. When we break that pact, things get ugly. Progress is not just about building more *stuff*: taller buildings, longer bridges, faster computers. It is about harnessing science and technology to make people's lives better. It is about using human ingenuity to overcome the limitations of the world around us. One of those limitations is human nature itself. We can be petty, competitive, and vain—or we can be generous, collaborative, and insightfully honest. It's up to us to choose progress (the real kind of progress) and strive toward it in all our actions, big and small.

That, in the end, is what I think my dad's THANKS sign was really all about. It's something we should all be flashing to one another, all the time.

CHAPTER 8

Why the Bow Tie?

As I walk through life, I often look down at people's shoes. No disrespect intended. I'm not trying to avoid eye contact with you. I'm paying attention to the imperfections in the world and looking for ways to help fix a few of them. You see, shoelaces are not just shoelaces when you view them through the filter of everything all at once. They are the raw material of knots, and knots are the embodiment of mathematical beauty; mathematical beauty is a fabulously useful tool for rational problem solving; and rational problem solving is, of course, the most powerful tool for changing the world. In my Nye's-eye view of the world, tying a well-crafted knot is like a personal promise to engage in that whole glorious process. I often have three such knots with me: two on my shoes and one around my neck in the form of my beloved bow tie. But when I look at the knots all around me—well, it's troubling. There's a lot of work to be done.

Try looking down yourself, and what do you see? Around half of the people I meet tie their shoes with bow knots that are prone to coming untied from the day-to-day flexing inherent in walking. These bow knotters often compensate by tying their laces with doubled knots, piling one asymmetrical knot upon the other in a desperate bid to keep it together—or worse, they repeatedly walk with loose laces dragging. It doesn't have to be this way. With a little more thought and attention, you can bring inspirational order to what may seem like one of the most mundane objects in your daily life. Plus, your shoes will fit better and stay tied.

Let's start with a simple experiment we can do together, right here and right now, using only the loosened laces on your shoe. Begin by tying one of the most useful of all knots, the square knot. It's also called a "reef knot," as it was and is, from time to time, used to reduce the sail area of a sail on a boat, to *reef* the sail in a storm or strong wind.

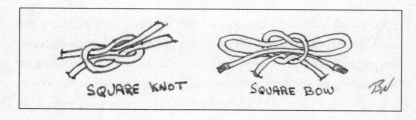

Wrap one lace over the other, then the second lace over the first one. You may have heard the expression "right over left, left over right." Look at that knot. It's beautiful, symmetrical; it's the marriage of two curves. This square, or reef, knot is square; I mean that it's symmetrical. It's the basis for the knot we call a "bow." Now, untie the second of the two wraps. You might go, "right over right, right over right" again. Please examine this knot. I hope you notice it's not as good-looking as the reef knot described above.

If you're like me, you might at this point exclaim, "Oh, the asymmetry!" This lack of balance found in about half of all conventional shoelace knots is heartbreaking. What we want in a square, or reef, knot is symmetry. Here, mathematical beauty is a means to an end. It's more than beauty for beauty's sake, although that ain't bad. It's a matter of function: A shoe tied with a reef knot will stay tied long after other, sloppier knots have come unraveled. In shoelaces, as in so much of physics, symmetry is the key to balance and stability.

When you tie a conventional bow on your shoe, check to see if its two loops, or "bunny ears," lie perpendicular across your foot, left to right, or lengthwise along your foot, toe to heel. If the loops (or ears) come to rest in a neat left-to-right position ("athwart," as we say at sea), that's the way we want it. That's symmetrical, and that arrangement will seldom come untied. This is what I call a "square bow." If you gently pull the loops so that the loose ends of the laces pop free, the knot that is left there underneath is the beautiful square knot. Even if you perceive your laces to be woven from slippery stuff, the square-bow knot will hold its own once it is gently but snuggly tightened. Or as the saying goes, any knot has to be properly "dressed." (For you crossword puzzlers out there, the loop of the lace is called a "bight." It's pronounced just like our word "bite," and it works wonderfully in Scrabble.) The unsymmetrical knot, on the other hand, will slip with each step. It will start to lose its shape, its integrity, and its stability the moment you start walking and put stress on it. Oh, the trauma; oh, the suffering.

As you may have inferred, I tie my bows by forming a single bight and wrapping the other end of the lace around the base of the bight. If you are among those who tie laces by finishing the knot with two loops, or bunny ears, it all works the same way. The bunny ears are your knot-ty-er bights. Allow me to reassure you bunny-ear, double-bight people: You can create a square bow just fine. If you tie the base over-hand knot, then form your two bunny-ear bights, and tie them in the opposite direction from your base overhand knot, you will produce a lovely square bow.

Now, I loved my grandmothers. They were both remarkable peo-ple. They raised my parents, after all, and I believe anyone who met either of them would say, "That girl has plenty of common sense." Nevertheless, the asymmetrical, not-quite-a-proper reef knot is, by long tradition, called a "granny knot." Sorry, Nana. Sorry, Mini. We seek a square bow rather than a "granny bow." If you have suffered lo the many years of your life with asymmetrical granny bows, you'll find it's a hard habit to break. But it can be done. Try this: Reverse the first wrap of your laces. Instead of going right over left, reverse that and go left over right. Then let your muscle memory take over for fin-ishing the bow, either by wrapping individual laces or by wrapping bunny-ear loops.

All this talk of shoe laces may seem like an unimportant detail of everyday life, but it is always underfoot—or literally atop foot. A shoe-lace knot is also a metaphor for the scientific approach to problem-solving. Too many people learned to tie bow knots in their shoes and accepted that imperfect, unsymmetrical, time-consuming route rather than dig deeper for a better long-term approach. So when I wax poetic about the beauty of a square knot, it's not only because I like showing off my sailor skills; it's because good design should be good all the way down to the details, even when we're talking about something fairly straightforward like tying knots. I think we should all make a habit out of expecting the best problem-solving from ourselves, and there's no better place to start than with design problems we encounter every day. That's where things like shoelaces work well or . . . not. (Get it? Or knot? Uh . . . sorry.)

There is another big idea in here, masquerading as a small one. Even if you have tied your laces the other way, in granny-bow fashion, for years on end, you still have a chance to change. This ongoing potential for improvement is at the heart of the scientific way of looking at the world. In politics or religion, changing your ideas can be risky or even heretical. In science, abandoning a decades-old habit in response to new information reflects a vital quality of open-mindedness. Such open-mindedness is essential for making a fundamental discovery . . . or for keeping your shoes tied.

Now on to the more personal side of knot theory, the one that is so close to my heart—just a little up and to the right, to be precise. I'm talking, of course, at last, about the bow tie. It is a symmetrical bow exactly, and I mean *exactly*, like the knots on your shoes. There are loops and tips, or bights and ends. And whether you're working aboard a ship, tying your shoes, or getting ready for a formal event, the loose ends are called "live ends." By the way, the small plastic or metal tips of shoelaces are called "aglets." It's another word for your Scrabble collection. And by the further way, the "bitter end" of a rope or line is the inboard end, which is often attached to a "bitt," a long wooden peg. (Trivia nourishes a nerd mind.) A bow tie follows the same mathematical rules and the same principles of symmetry that I have described for other kinds of knots. You have to form a bow around your neck, work the live ends, dress the bights, and snug it tight, and then it will look good. I mean, *great*.

As Jerry Seinfeld told me and a few other comics over brunch one Sunday, back when he was touring cities on his way up to the big time, "You want to dress better than your audience." Like proper table manners, paying some extra attention to your choices in cutlery or clothing shows respect for the people around you. That's another important tool for changing the world, and it's a beneficial way to go through life in general. I have found that when I dress up, I stand a little straighter. I project a version of myself that has the confidence and polish I feel on my best days, which allows me to have more best days. I feel more respect for myself, not least for knowing that I am carrying around with me a tidy showcase of applied mathematics: a skill that is as useful and essential as securing those shoes on your feet.

There's an old family holiday photo that shows my 4-year-old self sporting a very narrow and stylish bow tie, but my serious focus on the bow tie did not really get under way until I was a junior in high school. At the annual Girl's Athletic Banquet, the boys from the school served as the waiters. I figured further that if we were going to be waiters, we should make a good first impression by dressing like *professional* waiters. It occurred to me that if we were presenting the girls with their meals—especially their desserts—it was almost certain that some of them would have to talk to us, if only by accident, and if only for a few seconds. I figured it could be the start of something with a certain young woman.

My father, who was thoroughly skilled with knots, taught me how to tie a bow tie by demonstrating with the tie wrapped around my leg. It's easier than tying it around your neck, he explained, at least for your first few tries—or ties. Thighs and necks (of humans) are about the same distance around, so a leg-circling bow tie will come out very close to the right length. By practicing this way over and over while an episode of the *Perry Mason* television show played in the background, I acquired not only the ability to tie my tie but also an intuitive understanding of the mechanics and relationship of the loops (bights) and the live ends. By the time I made it through a week of shows, I could effortlessly knot up bow ties on others' necks, as well. So in the boys' room of my high school, I made the rounds and tied everybody's ties as we prepped for the Girl's Athletic Banquet. Then we served up some excellent school-style repasts for the ladies. And it worked—sort of. The girl of my dreams really did speak with me but, well, it wasn't a deal-closer. But I learned a lesson that was more important (in the long run) than closing some sort of adolescent deal: building my confidence. With a bow tie round your neck, you make a respectful first impression. That part stuck with me.

During this experience, I discovered another important functional advantage of a bow tie: It does not hang down over the buttons of your shirt the way a straight (or long) tie does. This feature prevents the tie from slipping into your soup when dining, or finding its way onto a server's tray while serving, or flopping into your flask while you are swirling surfactants or solvents. A bow tie wraps up fashion and function in one tidy package. It's all good, is all I'm saying.

Nevertheless, I switched back to more conventional neckwear for a while. As I moved up through school and on through life, I generally wore a straight tie to work or church. It's what I felt the Man (or the Woman) expected me to wear. But in the 1980s, shortly after I started giving stand-up comedy a try, I experimented with bow ties as a way to set myself apart from the dozens of other would-be stand-up comics and as a way to keep from having my tie get in the way of the physical aspects of my comedy. I would flail my arms and produce balloons, or the occasional wrench strapped to my leg. I was, I admit, not especially funny. Now and then I would get laughs. But the reaction to my gags aside, I became more and more comfortable performing in a bow tie. It became part of the act, part of what I looked like on stage. A feature of stand-up comedy generally is that it has to be honest. The character one presents has to be self-consistent; it's hard to laugh when the performer is not authentic. There's an old saying in the theater: "You can pretend to be serious, but you can't pretend to be funny." I obviously had an edge there because I started out not just funny but funny-looking, as well.

After work, I would come home, take a nap, remove my straight "this one's for the Man" tie, then head out to one or another comedy club. Wait—I would take off my tie before I took a nap . . . Upon awakening, I'd put on a bow tie and head out. During all this, whether it was a straight tie or a bow tie, I also wore a shirt. I was just working with our society's "rules." On the very first *Science Guy* segment that I did on television, in January 1987, I wore a straight tie there, too. I was on TV, trying to fit in once more. As the weeks went on that spring and I produced and performed more *Science Guy* bits, though, I found that a bow tie is just more practical. No slipping or flopping in soup or solvents. No blending in with the crowd.

Oh, there was a little more backsliding. I experimented with straight ties again around 2004 and 2005, when I was doing another show called *The Eyes of Nye*. It gave the viewer a new perspective on science-related issues, specifically ones that do not have a straightforward answer. Can we create new antibiotics that won't produce resistant pathogens? Can nuclear waste be safely stored? Why does any organism bother with sex, etc.? Since I was striking out in a new direction, the producers and I tried me in a straight tie. It was okay, but just okay, because at this point I had been wearing bow ties for years on my show as the Science Guy. And, by the way, I feel that a big part of the success of *The Science Guy* show was and is that I was pretty much myself on there. With Bill, "what you see is what you get," our editor Felicity used to say. I think she meant it as a compliment.

I had come to prefer the bow for its practicality and distinctiveness, but there's still more to it. The bow tie also comes with a rich history as an artistic type of knot—another dose of delightful nerd trivia. The tradition of wearing a tie caught on among 17th-century Croatian mercenaries during warfare. Combatants would don a scarf so they could tell what team they were on while working to maim, decapitate, or otherwise kill their enemies (the other team). But soldiers are not always in combat; hence scarves found their way into the military dress uniform of the time, as well. During the 18th century, French aristocrats caught on to the trend, and the Croatian scarf evolved into the cravat (worn by Croats), made of fancy cloth and knotted according to rules unfamiliar to regular middle- and lower-class folk. The cravat, in turn, evolved into the bow tie, and here we are. One thing I especially like about this story is the way that tie-wearing transformed from a preparation for war into a peaceful expression of respect for your allies and peers. Yet another step in making a better world with knots!

At any rate, bow ties are my signature now. It would be difficult to turn back. When I go to college campuses these days, bow ties have become a theme. A lot of the students who show up wear bow ties of some kind. It warms my heart. Like the pocket protector or the slide rule, the bow tie has evolved into a nerd badge of honor. It displays both a mixture of deep respect for tradition and an easy comfort with standing apart from the crowd. Being open-minded and a team player are generally regarded as good attributes of a well-rounded productive employee or boss. Bow tie knots may help you think in new productive ways. They also show a willingness to stand apart and prioritize function and design in your everyday life—guiding aspects of the nerd lifestyle. Am I overdoing the bow tie bit? Is such a thing even possible? Clearly knot (not . . . uh, sorry again).

A well-tied knot, and especially a well-tied bow tie knot, signifies an appreciation for timeless symmetry over the ever-shifting whimsies of commercial style. I'm encouraged, then, that bow ties have lately become downright fashionable again. I like to think that my persistent resistance to the latest trends, and that of my like-minded nerds, helped nudge the bow tie back. With the return of the bow tie has come renewed demand for the practical knot theory that allows it to look so crisp and proper. The nerds have a leg up here, so to speak. According to accounts from various cocktail-style events—I hear things, people— many folks today don't know how to tie a bow tie, and they feel self-

HERE'S A LITTLE DESIGN GIFT FOR YOU, DEAR READER: a guide to several of the most useful knots for when you're bringing home a mattress or holiday tree atop your car, securing a boat to the dock, getting your dog leashed reliably while you order a cup of coffee, or just being confident with a string in your hand. If you don't know these knots already, I hope you will challenge yourself and try them out. Even if you have no car, no boat, no dog, or no shoes, it's good to develop new skills every once in a while. It's a simple matter of knowledge as power. Every knot is also its own little lesson in symmetry and the distribution of forces—a microcosm of mathematical elegance. I think everyone should know how to tie the following:

SQUARE KNOT   BOWLINE
SQUARE BOW   CLOVE HITCH
TWO HALF-HITCHES   RIGGER'S HITCH
                   (TRUCKER'S HITCH)

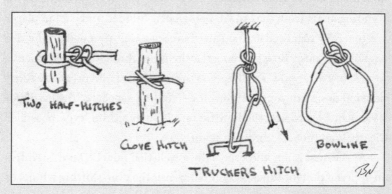

conscious about wearing one. When I'm at a grown-up event with a bunch of adults in attendance, it can happen that I'm the only guy there wearing a proper bow tie.

It could happen to you, too. My feeling is: C'mon people—take a chance. Although one cannot unbutton one's collar as easily as one can while wearing a straight tie, the bow tie has a certain knotty appeal that is difficult to achieve with a straight tie. Try it. Tie it. I dare ya.

With those half-dozen knots, you can tie most things to most other things. Most people will find they can get by nicely with that set and no others. Once you come to appreciate the beauty of a well-constructed knot, though, you may discover that the knowledge is addictive (in the good way). Like mathematical equations, knots come in almost endless variations that may look superficially similar but have wildly different properties. Some are nearly impossible to untie, others look strong but easily slip apart, still others will easily but nearly unbreakably join two separate pieces of cord. If you are like me, you'll want to keep going and get to know some of the other famous knots.

SHEEPSHANK
MAN-O-WAR SHEEPSHANK
DOUBLE BOWLINE

BOWLINE-ON-A-BIGHT
SPANISH BOWLINE
(DOUBLE FORKED BOWLINE)

I included those last few knots especially for the hard-core nerds among us. Most people will never learn them, but each one is important and beautiful in its own way. Each has a different origin, a different history, a different utility. The double bowline, for instance, is particularly useful in rock climbing. The Spanish bowline not only is a truly elegant work of double-looped symmetry but also turns out to be really useful for lifting a body. If you fall into a cave or crevasse, your rescuers might lower one down to you. Put your legs through the loops, hold on to the cord, and you are on your way back to safety. It's an exquisite solution to a terrifying problem. Welcome to another corner of the everything-all-at-once world.

CHAPTER 9

Land of the Free, Home of the Nerds

As much as I try to think globally and universally, there's no escaping my local origins. I was born in the United States, got my engineering degree and engineering license here, work here, and live here. It's possible, then, that I'm not entirely objective about the quality and effectiveness of the American government. Be that as it may, I am amazed, humbled, and filled with reverence whenever I visit the US National Archives in Washington, DC. It is located downtown beside the National Mall, right near several far more famous landmarks—including the Smithsonian Institution's National Museum of Natural History, National Museum of American History, National Museum of African American History and Culture, and the wonderful National Air and Space Museum—so the National Archives are often unjustly overlooked. I was just there and, wow, it knocked me out all over again.

No election result can change the way I feel there. If you're ever in DC, I strongly urge you to visit this place.

To me, the National Archives is nerd nirvana, not just for Americans but for anyone who wants to see what can happen when you put scientific principles to work in building an entire nation from scratch. Here you can view the letters and documents recording that process as it unfolded. The founders of this country were products of the Enlightenment, the 18th-century intellectual movement that regarded reason as the highest quality of the human mind. Thomas Jefferson, Benjamin Franklin, and company sought to create a government that was better than any that had come before, and they believed that the way to do it was to ask honest questions, examine evidence, carefully debate the pros and cons of various solutions, and engage in rational action. "Everything all at once" is not an idea that began with me, not by a long shot. It is woven into the fabric of this nation, even if people seem to forget about that a lot of the time. Revisiting that spirit can be a constant source of inspiration, no matter where you live in the world.

As a visitor to the National Archives, you begin by ascending a great many steps; it is an architect's way to affect your senses. Climbing the wide stone stairs metaphorically (and literally) elevates your mind along with your brain. The building itself has a stately marble form, reflecting the United States' long appreciation and emulation of ancient Greek columns and edifices, as well as the influential Greek experiments in representative democracy. In the main gallery, you'll find a remarkable collection of documents. The museum's displays change all the time, so you will see something different with each visit. When I visited, that collection was themed around the topic of legal rights, and it included copies of both the USA PATRIOT Act and the

Magna Carta. Those rotating exhibits are a stirring reminder that the National Archives is, at heart, an enormous data-storage vault.

As you make your way toward the back of the great hall, you'll catch glimpses of dimly lit sheets of parchment. They are dramatically large, almost as big as a typical poster for a blockbuster movie or a bucolic National Park hotel, cluing you in about their historical importance. The first time I saw those documents, I was a young kid, not yet capable of comprehending what I was seeing. But my second encounter is etched in my memory. I was a teenager, and when I walked into the Archives, I remember thinking, "Hmmm, that looks just like the Declaration of Independence, and that other thing right next to it looks just like the US Constitution." It took me several moments to grasp that I was seeing the real things—not prints or facsimiles, but the bona fide founding documents of the United States. Right there before me was Thomas Jefferson's handwriting, setting forth a series of enormous ideas about freedom and independence on a very thin sheet of parchment.

Talk about changing the world! The Declaration of Independence and the Constitution (along with the Bill of Rights—it's right there in the National Archives, too) describe the principles for creating not just a new nation but a new kind of government. Jefferson, Franklin, James Madison, Alexander Hamilton, Gouverneur Morris, John Hancock, and the other founders mined history and contemporary political philosophy for the best concepts about how government should work. They took inspiration from the dreams that had already led generations of settlers to the New World: freedom from tyranny, fair representation, an escape from superstition and repression, and an ability to observe any kind of religious faith, even no religion at all. The separation of church and state was central to the idea of a thoroughly modern nation that would embody the Enlightenment ideals of liberty and reason.

The potential for progressive change that's built into the United States Constitution is reminiscent of Darwinian evolution and, more fundamentally, of the scientific method itself. Just as science doesn't claim to attain absolute truth, the Constitution does not claim to achieve the utopian ideal of government. It promises "a more perfect union" not "a perfect union." The founders understood that the documents they wrote were not going to include the last word (or words) about how to run a just and peaceful society. Instead, they acknowledged that the nation's laws had to allow for change in response to new needs and new information, just as scientific theories change in response to new ideas and new data. Contrast that with the ways of a monarchy, in which a king can make laws unchallenged and order any action, including the lethal decision to go to war. Monarchy is inherently static—unless the people rise up in revolution. The American Revolution was, in a sense, a scientific revolution as well as a political one. Even with the antiscience forces emerging with the 2016 election, the system of government will adapt; change and a process of governance akin to the scientific method is built in.

When you read the letters of Jefferson, Franklin, Hamilton, and the other founders, you see that the writing is florid, impassioned, full of a sense of gravity. I marvel at the risks those guys took. If the Revolutionary War didn't turn out well, they all would have been shot or hanged or beheaded, depending on the winning general's whim for what they were attempting. They accepted that what they were doing was more important even than their own lives.

Just imagine the mood in the early summer of 1776, when breaking free from England changed from an abstract threat to a world-changing reality. Jefferson and the other authors of the Declaration of Independence could have crafted a document simply formalizing a military rebellion—and in part that's what they did, listing the reasons for wanting to be rid of King George III. But they went much further. They were thinking, "If we could design this thing top-down, we could have a system that continually refines itself, that continually sharpens its edge; we would have a fantastic government." Their defining sentence is so famous that it's easy to forget how radical it was at the time: "We hold these truths to be self-evident, that all men are created equal, that they are endowed by their Creator with certain unalienable Rights, that among these are Life, Liberty and the pursuit of Happiness."

The greatest intellectuals of the American colonies had gathered with a giddy sense of purpose. They were declaring independence not just from a king but from an entire system of thought. They were declaring that everyone has equal rights under the law—at least as far as they were able to extend that idea within the attitudes of the age. Equality remains one of the greatest goals for political progress around the world. The founders of the United States were still human, of course, working within the assumptions, biases, and ideologies of the day. Women were pretty much excluded from the Constitution when it was written. Black people were discussed as though they were two-thirds of Caucasian guys and gals for the census. Worse, they were treated as nonentities—property rather than people—when it came to voting and other basic democratic rights.

In nerdy terms, you might say that the founders were creating the best system they could using only the information available at the time. Doctors of that period did their best to heal, even though they did not know very much about the germ theory of disease. Engineers produced helpful technologies, hindered as they were by the lack of a complete understanding of the laws of thermodynamics or the behavior of atoms. We don't normally think of laws as things that people discover, but that is very much what was happening during and immediately after the American Revolution. A group of scholars pulled together the best examples of good governance that they could, limited though they were, and then sought to discover something even better. The founders were not all of one mind. They disagreed, fought, compromised, grudgingly accepted that a compromised achievement is better than an unattainable perfection. They went through all the trials associated with everything-all-at-once thinking and ended up with something marvelous as a result. Still flawed, of course; how else could it be? But the central ideas behind the Declaration of Independence and the United States Constitution were hugely powerful and important.

From a scientific point of view, people are all just people. We are a single subspecies, *Homo sapiens sapiens*, with remarkably little genetic diversity. We are all descended from a common ancestor, and the regional variations that we call "race" are minuscule compared with our overall biological and genetic identity. Centuries before these ideas were established scientifically, Jefferson and the other founders established the same basic concept politically. In their system, nobody gets to be a queen just because there wasn't a suitable man around, or because she married the right man. Nobody gets to be a king just because his father was a king. "All men are created equal" is the aspirational promise of a data-driven world in which people are judged on their actions, not the uncontrollable social circumstances of their birth.

At this point you might be thinking, geez Bill, aren't you getting a bit carried away with the "USA! USA!" and the "it's all science!" angles here? I mean, it's not like the founders were actual scientists. Ahhh, but—actually, that's exactly what they were. I ain't braggin' on my people, but the guys we call the Founding Fathers were, almost to a man, "natural philosophers," which is what we now refer to as "scientists." They studied the way governments had been set up before, treating history as their laboratory. They thought deeply about how to construct a system that would keep running smoothly and fairly, long after they were gone. They also studied the workings of the natural world. It's not just Benjamin Franklin and his electricity experiments with glass rods and a hypothetical kite. He helped map the Gulf Stream. He invented bifocals, the lightning rod, a high-efficiency stove, an improved electric battery, and . . . um . . . a urinary catheter. Thomas Jefferson designed an improved agricultural plow, developed techniques for excavating archaeological sites, and published the first paper on paleontology in the United States. To me, it's not a stretch to call this group the Founding Nerds.

I feel a distinctly personal connection to them and their science-y ways. I can trace my family line back to Benjamin Nye, who set up shop in Massachusetts in 1656. (I wasn't kidding about my local origins.) He left England searching for adventure, as well as for an escape from the intrusive commercial rules set up by the Church of England. In Scandinavia, *nye* means "new," and indeed the Nyes were newcomers. First they left Denmark for England, and then they left England for the New World, restlessly seeking a better life each time. My ancestors participated in the Revolutionary War. They were seafaring and business people, adventurers who were always looking for the next thing to explore or invent. I'm one link in a long chain. My dad the tinkerer called himself Ned Nye, Boy Scientist. And my mom was "all that" when it came to chemistry and cryptography; she taught me to tackle puzzles with the same intensity that she went after World War II codes. So here I am fighting the old fight, some 240 years after Benjamin Nye and a couple of million other colonialists set this nation on its present course.

Before I wander too much further, I want to say that I am very aware of our continent's first peoples: those who walked or boated to North America in ancient days from Asia and were brutally forced from their ancestral homes by European settlers centuries later. It became a war, and as in too many wars, there were countless humanitarian crimes committed. It is difficult to compensate the victims fully. Despite the atrocities perpetrated both before and after the founding of the United States, I maintain that this nation's government was (and is) a remarkable experiment, one that created mechanisms for moral as well as political process.

As messed-up as my native country seems sometimes, people still come here from all over the world, and many more yearn to do so. They dream of working in the United States and of becoming American citizens. A big part of what makes them want to take their chances in a strange land is the fairness. The laws in this country recognize merit over pedigree. There is a kind of nerd honesty that was incorporated into the system from the start, and the United States still exemplifies it. The Constitution was intended to establish a government better and more humane than any that had come before. This is another key Enlightenment idea enshrined in the founding documents that I communed with at the National Archives: History progresses toward better conditions, and rational thought is what propels that process forward.

Nerds treasure knowledge because it is what allows us to find answers and develop new solutions. It promises that tomorrow will be better than today—because we will make it so. That attitude is inherently progressive; and in that sense, there is no question in my mind that Jefferson, Franklin, Hamilton, and company were full-on nerdy progressives, too.

❯ ❯ ❯

Progress can be slow, but with time the consequences grow huge. That's the way of any system that is built on a consistent, rational set of rules that allow ideas to compete and adapt. Just think how different the United States is today from what it was in 1789. The abolition of slavery was the most wrenching advance, but it was hardly the only transformative one. My great-grandmother raised a lot of eyebrows in my family in 1913 when she chose to skip attending to the needs of her 2-day-old first grandson (my uncle) and instead march in a suffragette parade, demonstrating for women to get the right to vote. She was outraged that half the population would be excluded from the chance to guide the government. I'm outraged, too, when I think back on it. But on another level, I think, "How cool is that?" Getting her place at the table of governance was more important than . . . well, almost anything else in her life. She finally got her wish in 1920 with the ratification of the 19th Amendment.

I've seen democratic progress in my own lifetime, as well. When I was a kid, people in my hometown of Washington, DC, could not vote in presidential elections. It was a weird relic from the establishment of Washington as a federal jurisdiction, separate from every state. Keeping the people there out of presidential politics must have seemed like a good idea to those constitutional founders. "This town will be set aside, and people who work here will not be swayed by having to pick sides in the most important vote available to citizens of their country," they might have reasoned. "And besides, we're only talking about a few dozen hundred people." But Washington grew to hundreds of thousands of people, and the arrangement became obviously unfair—so the law was changed. I was a kid, but I remember it clearly. My parents, both veterans of World War II, voted for president for the first time in 1964. That was just 1 year before the Voting Rights Act, another milestone in expanding and protecting American democracy. (Washington still doesn't have full congressional representation; change takes time.)

Such changes are possible because of the guys who crafted the documents hanging behind the thick, bombproof glass in the main hall of the National Archives. The authors of those documents were thinking deeply in the style of everything all at once. They knew that they were at the very beginning; they were creating from whole cloth or whole parchment, making the critical decisions that would shape everything that followed. They knew that they had to be utterly clear about their goals and relentlessly honest in considering all the potential consequences of their actions. They drew from the best-known intellectual sources (from Aristotle and Plato to Francis Bacon and John Locke) to find ways to reconcile a nation's needs for strength and justice. They realized that they had a shot at making a big and lasting change in one government and in the nature of government itself. The musical *Hamilton* captures this spirit in one of its most memorable lines, when Alexander Hamilton passionately vows, "I'm not throwing away my *shot*!" He had a shot to change the world, and he took it.

To my mind, the ongoing process of American political change is reminiscent of biological evolution. That's what I was driving at before when I described our legal system as something that allows ideas to *compete* and *adapt*. The people who crafted the Constitution and the Bill of Rights realized that it was not enough to believe in the possibility of progress; they needed specific institutional mechanisms for enabling it. Drawing on Enlightenment philosophy, they came up with an approach to government that had intriguing parallels with Charles Darwin's theory of evolution by natural selection—and they did it a decade before Darwin was even born.

The founders conceived a legal framework based on certain inviolable rules of liberty, tranquility, justice, and personal welfare, but they also allowed laws to adapt from the bottom up based on the will of voters. For example, we decided to drive on the right side of the street instead of the left. It's in our laws, but that kind of detail is not in our Constitution. The president can direct the actions of the government, Congress can continually write new laws, and the justices of the Supreme Court can continually interpret the Constitution in the light of new social, political, economic, and scientific realities. In nature, better-suited organisms outcompete the less-suited ones. In a constitutional democracy, good laws can outcompete the bad ones or the less-good ones. The Constitution sets the ground rules, but it is the beginning, not the end, of the American legal system. From there, you can get innovations such as expanded voting rights, broader freedoms, and new governmental structures like, say, the National Science Foundation or the EPA. I'm amazed by all this reason-based foresight every time I visit the great hall of the National Archives.

Whether or not you make a pilgrimage to the National Archives—whether or not you are a US citizen or resident, for that matter—you can witness the legacy of that revolution. You do it all the time, because it is all around you in our roads, our power lines, our environmental regulations, and our constitutional protections. If you live in another country, you've still been touched by those powerful Enlightenment ideas, as well. The founding concepts of democracy, equality, and a rational approach to law are built into the freest and most productive political systems around the world. There is a deeply scientific way of thinking underlying those concepts.

Whenever I work on a difficult project, I try my very best to resist thinking that "it's done" as soon as it feels like I have reached the nominal end. Whether it's an engineering sketch for a portable baseball pitcher's mound or the script for an "audience connection" piece on my Netflix show, *Bill Nye Saves the World*, I recognize that the first-draft versions of things will almost always need more work. I strongly discourage the writers, producers, or editors from using the word "final" at the top of any document. Nothing is final on a TV show until it's recorded and on the air. The same is true of any creative project. If you want to do good works, you have to build in the ability to change and adapt. Nature does it; so should we.

To me, the most moving thing about staring at those very old pieces of parchment is remembering how utterly modern they are. Sure, they are illuminated dimly to preserve their fading inks, and they are sealed in a vault of inert argon gas to keep the paper from turning to dust, but the words on them are as alive as ever. They are stirring reminders that in law, as in science, good-enough thinking is never good enough. That's why the Constitution needs constant reinterpretation. It comes back to that wonderful notion of progress. Maybe we'll even get rid of the Electoral College one of these days. You wouldn't want to base modern medicine on ideas from the time before the germ theory of disease, or try to build a jet engine using the old theory that heat is a fluid called "caloric." Fortunately, the founders embraced a never-ending search for better ideas and better solutions. The Constitution is one of the most beautiful examples of everything-all-at-once thinking, and it's a personal inspiration as well as a political one.

Say what you will about the shortcomings of American society, and there's certainly a lot to say, but our progressive style of doing things is a constant reminder that such greater things are possible. We do not have to accept things the way they are. We can nudge, adapt, revise, evolve, improve . . . and change the world.

CHAPTER 10

Everybody Knows Something You Don't

Carefully consider the title of this chapter. I hope you can take a moment and let those words really sink in. It is such a simple idea, yet I find it constantly humbling and empowering. You are surrounded by people who are experts in one field or another, and most of those people are ready to share their knowledge freely with you. All you have to do is approach them openly, generously, and attentively.

The expertise that others have might not be what you or I expect. It might not even be something you recognize immediately. I can pretty much guarantee that a janitor knows more about chemical interactions among cleaning solvents than the average person. A flight attendant is sure to be trained in the latest CPR techniques. A grocery checkout worker can probably tell you the exact texture of a ripe mango or the true meaning of the sell-by date on a carton of milk. It's marvelous to think about all the knowledge that's around us at every moment.

But in order to tap into all this great wisdom, you still have to reach out and ask for it. There's no shortage of reasons why we don't. We are too busy, too shy, too proud, too oblivious. A lot of the time we don't even notice other people—or we are so certain that our own knowledge must be superior (or at least plenty good enough) that we don't bother to ask. But think about it: No matter how many things you've studied or how many life experiences you've had, somebody else has studied things you haven't and has experienced things you haven't.

I'm a walking case study in knowing a lot of stuff about a lot of stuff and still having a warehouse full of ignorance. I especially had that feeling when I was first starting out as an engineer at Boeing in 1977, at a time when I had more education than life experience. I was about to work on engineering upgrades for the company's biggest and most famous jet, the 747. Fortunately, my boss, Jeff Summitt, was an amazing guy. He grew up with slide rules, and he had an intuition about forces, pressures, and mechanisms. I wanted to impress him with my competence and tried to emulate him so much that I even wore the same brand of safety shoes he wore and ordered the same thing he did for lunch in the cafeteria.

There was one Boeing test pilot who was really bothered by vibrations in the control yoke—the pilots' steering wheel—on the 747. I say "vibrations," but what he picked up was more of a subtle buzz. You had to concentrate for a moment to feel it. But once felt, you noticed it every time you touched the yoke (which a pilot does all through a flight). Jeff assigned me to figure out how to get rid of the bad vibes. The Boeing 747 was the first commercial airliner that was all fly-by-hydraulic, meaning that there were no direct, unpowered connections between the pilots' controls and the aerodynamic surfaces (the ailerons, rudders, and elevators) that steer the plane. To give the pilot a sense of what the plane is doing in the air, there was a gizmo called the "feel computer." Using hydraulic fluid at a pressure of 2 megapascals (3,000 pounds per square inch), the loaf-of-bread-size computer generated artificial control forces. Just as the steering wheel of your car pushes back on your hands when you turn, the feel computer provided force-feedback in the pilots' yokes in the cockpit. The 747 has four redundant hydraulic systems to deal with the extraordinary forces needed to control such a large aircraft flying at such high speeds. That arrangement means there is almost no chance that a single fault, or even a combination of faults, can disable the plane.

So what was causing the vibration? A 747 is a complicated piece of machinery, and Jeff helped me break down the problem. With all four hydraulic systems running at full pressure (i.e., normally), they sometimes work, or at least push, against one another a bit. When I first looked at the problem, Jeff dryly pointed out that any vibration the pilots were feeling had to be coming from those hydraulic systems, not from the air. I investigated further, examining the entire system piece by piece. I finally concluded that the tubes themselves were to blame; along their considerable length, they were ever so subtly creating a high-frequency wave in the hydraulic fluid. Nothing to see, just a buzz in your fingers.

Now we had a better-defined problem: how to dampen the artificially generated vibrations. We added an extra length of hydraulic tubing and made the pressure wave destructively interfere with itself. It was a clever trick from wave theory. When the pressure wave went high, our new, artificially created antiwave went low and canceled out the whole problem. Since I was fresh out of engineering school, Jeff assigned me to do the math. I computed the speed of the waves in the fluid and the volume needed to generate the exact right antiwave. My numbers guided our design, and the vibration went away. I'll never get over how little I knew about the things Jeff had me doing. He not only had detailed knowledge about hydraulics and control systems but also knew how to filter his knowledge. Such filtering is what we typically call "intuition." It took me a while to appreciate just how much Jeff knew that I didn't, but he was patient with me. He never made me feel embarrassed about asking questions. He made me part of the team, treated everyone with respect, and encouraged us to share our best assets and strengths. We all contributed to the designing, building, and testing of our antivibration system. It was the beginning of a journey.

Years later, I left Boeing and worked a number of other engineering jobs. In 1982, I joined Sundstrand Data Control, the company that makes most of the black-box flight data recorders for airplanes. It's right across the highway from what is now the Microsoft campus. At the time, it was across from a cow pasture.

Along with those crash boxes, Sundstrand made all sorts of other aviation electronics, or avionics. Many of those avionics packages incorporated small, astonishingly accurate accelerometers—ultrasensitive devices that measure small pushes or pulls. When I say "ultrasensitive," I mean *ultra* ultra. Even back then, the accelerometers were so precise that they could measure the Moon's gravity from the surface of the Earth; that's a pull of about 30 micro-gees of acceleration. To put that in more concrete terms, my home scale says that my weight on the surface of our planet is a lean, mean 70 kilograms. When the Moon passes overhead, its gravity pulls me slightly and I weigh a bit less. Precisely speaking, if I weighed 70 kilograms on the scale before, my weight decreases by 30 parts per million to read 69.99979 kilograms.

No, you cannot notice a tiny tug like that, but the *ultra*sensitive autopilot in a jet airliner most definitely can. Instead of measuring the Moon, the accelerometers in autopilots sense the effects of the pulls and pushes of the airplane's motion in what we call "inertial space."

They can also be used underground to help guide the drill in a mine-shaft or an oil-well drill casing. At Sundstrand, the specific problem my fellow engineers and I were working on was how to find, with extremely high precision, which direction pointed straight down anywhere a drill was working underground. You couldn't see up and down from deep inside the earth, of course, and yet you had to know precisely where you were pointing your drill bit. We were using a combination of accelerometers in x-, y-, z-fashion (basically east-west, north-south, up-down) to aim and steer complex heavy drilling systems.

Here was another problem whose solution exceeded the extent of my knowledge. When you're designing a system like this accelerometer setup, you have a great many choices available to you. For instance, you could design some of the parts to exquisitely tight tolerances by building them in a special temperature-controlled room. That entails a lot of work, but it may allow you to make another part to not-quite-so-precise tolerances, which in the end could simplify the overall assembly. Alternatively, you could make all the parts with somewhat loose tolerances, but then carefully align the parts and secure them in precisely the right locations using an additional fixture that is built to extremely tight tolerances. That way, you're letting the fixture do the work of aligning, smoothing, or assembling.

So which is the right solution? If you'd asked me then, as my team did, I'd have been stumped. The second option didn't even seem like a realistic possibility to me: "You can do that? Aren't alignments like that beyond the capability of our instruments?" That was my thinking at the time, but only because of the constraints of my own knowledge—not because there was no solution. I wasn't sure there was a way to solve the problem, and I had only a rough idea of where to even start. I was literally not sure which end should be up (and which down, and which north and which south, etc.). Sometimes you need help figuring out how to get past the constraints of your own knowledge. That's where I was at that moment. I wasn't quite panicked when faced with the assignment; let's go with "concerned." So I asked my very sharp designer coworker, Jack Morrow. He didn't know the right approach, either, but he knew *how to know*. He told me, "Go ask the guys in the shop."

Jack was referring to the machinists, the guys who cut metal using large, elaborate machines with superhard cutting bits, blades, and broaches capable of turning out virtually any part you could imagine. I didn't know about the shapes and tolerances that can be created and achieved by cutting metal, but Jack pointed out that the guys in the shop certainly did. Then he drove the point home with a disarmingly simple sentence that has stuck with me: "Everyone you'll ever meet, ever, knows something you don't."

I walked down the long hallway to the machinists' shop and spoke with Roger, Mose, and Phil. They showed me which parts of my would-be inclinometer (measures inclination using acceleration) would be easy to build and which parts would be hard. They talked me through the design process from their perspective, warning me about common mistakes that got made when setting expectations about the metals and clueing me in about useful tricks to avoid misalignments. They helped me, and in the process, helped the whole team design a better instrument. And you know what? They didn't laugh or make me feel dumb. They didn't joke about the new kid still learning his way around a jig-bore machine. They were happy and proud to share their expertise. I mean, who doesn't enjoy a chance to show off what he or she knows? I also got the impression that they appreciated being consulted early in the design, when their advice really made a difference. They got to be part of the design team from the start instead of being summoned in frustration at the end to fix problems locked into the final design by people who didn't know any better. It was a win-win for all of us and ended up saving everybody a lot of time as we worked together from the beginning to come up with an optimal, practical solution.

When I was sitting down with my boss at Boeing, it was obvious who had the expertise. What I realized in that machine shop a few years later was that a lot of the time the people who have the best knowledge are not so obvious. In this case, they were just steps away, but I didn't realize it. After a few schoolings by the machinists, I went back to the (actual) drawing board and incorporated the machinists' suggestions. I took their advice about what features to make datum planes, what dimensions to specify as the "basic" ones, and what parts to align. Under Jack's tutelage, I taught myself the then newly adopted standard called "true position dimensioning and tolerancing." If you don't know what these things are, join the club. Now you have an inkling of how I felt. There's a sense of being useless when you don't know what is going on. But then when you ask and learn—oh, what satisfaction.

I won't pretend that everyone had the same status at the company. There were sizable wage disparities between the machinists, who had gone through a long training for their exacting trade, and Karl, the floor sweeper. The softball teams usually were organized around the groups of people who worked closely together. Engineers with engineers, machinists with machinists. Nevertheless, we all had respect for each other. We all had a role to fill, and we understood that we could not do our jobs correctly without the cooperation and expertise of everyone else. In places where everyone is collaborating smoothly that way, often the thing you notice most is—nothing. You do your job, and things just seem a little easier than you expected because everyone is pulling together as a team. But I've also been in countless workplaces where people did not appreciate each other's expertise, and there's a big difference. It's frustrating, depressing even. There's no way to get great things done in those kinds of settings.

There are many ways to promote open listening and even to institutionalize the idea. Big corporations like Boeing and nonprofits like The Planetary Society have organizational charts listing the hierarchy of employees. At the top is the chief executive officer. Partway down, we find the middle managers. At the bottom, you might find Amy the engineer or Karl the custodian. But in the hotel business, it's the other way around. The CEO is at the bottom, and the hotel guest is at the top. It's a cool way of looking at things. At The Planetary Society, we've created a similar type of chart to highlight the members who support us. On my *Bill Nye Saves the World* Netflix show, the audience member is the most important person. On our show's call sheet (the document that lists everyone working that day and what they are doing), I had myself listed as part of the "On Camera Department" rather than the more traditional "Talent Department." My job is to deliver words and images for the audience, but I'm not the only one with talent. The camera operators, makeup artists, audio recordists, gaffers, and grips all know plenty of things I don't. Wow, do they . . . It's a team, and we work hard and collaboratively to make good shows. We don't need to fake it, because we can turn to one another for answers. We respect one another.

I like to think nerds are natural listeners, because nerds are natural learners, which gives them (us) a huge advantage in breaking through some of those personal barriers. Even so, it's a constant battle. In many ways, big and small, our society conditions us not to pay attention to people who are different from us. But you can't apply the everything-all-at-once method if you don't make yourself open to the "everything" part to begin with.

These ideas about the importance of expertise and respect came into focus for me after an incident last year when I was on a plane coming back from Guadalajara, Mexico. I had just attended the International Astronautical Congress, and watched as Elon Musk showed the world his plans for sending hundreds of passengers to Mars aboard enormous rocket ships. People there were (however briefly) likening Musk's presentation to John Kennedy's 1962 speech about going to the Moon. The audience was wound up; they were all buzzing about the Mars ideas on the outgoing plane flights home. Incidentally, a lot of that debate was centered on the theme of expertise: Did Musk really know how to pull off his fantastic-sounding plans, or had his dreams exceeded his abilities?

While these lively, and at times heated, conversations were going on, I was comfortably ensconced aboard a Boeing 737 and feeling my usual anticipation of ordering a beverage from the approaching flight attendant. She was methodically working her way down the aisle, taking care of the passengers with a seamless, well-studied cheerfulness— until the guy in front of me suddenly began ranting about how this particular flight attendant had "dissed" (disrespected) him. He claimed that he hadn't been asked what he wanted to drink before the person next to him, that the food options were terrible, that he did not deserve to be treated this way, etc.

I was incredulous. I was thinking, "Dude, get over yourself." I nearly lost my own temper in response, and I was a row behind the guy. I didn't actually have to deal with him. But my goodness, the flight attendant was amazing. She remained extraordinarily calm and tried out different ways to defuse him. She kindly explained that she was sorry the evening's meal selections were not to his liking. She asked if perhaps he'd like one of the official-issue snack boxes, which feature hummus and cheese along with olives and crackers. The guy kept going for another couple of minutes, but eventually her calmness did its work. Maybe her kindness finally got through and made him feel ridiculous. Maybe he finally decided he'd received the respect he deserved. Whatever it was, the flight attendant figured it out and made peace with him. I guess people like her are trained to be resolutely polite while handling unruly passengers, but it made quite an impression on me and on everyone else seated around the angry guy.

Fundamentally, I think this guy wasn't thinking of the flight attendant as a person with thoughts, feeling, and insights. He may be the kind of person who likes to have control over people in the service industry because he lacks control in other parts of his life. He may have been thinking that in that setting on the airplane, he could be in charge because he was a customer. But what I observed in those few awkward minutes was that the flight attendant was the one who managed him. Apparently he presumed that because she was working in a service job she was less than him and not deserving of respect or civility, but in truth, he was the one who lacked it. In his frame of mind, for whatever reason, he could not see the flight attendant as someone with a great deal of expertise, as a professional doing her job to a high standard.

That ugly scene drove home all the things the flight attendant must know. She knows about flight schedules, airplane operations, and the different technical styles of the pilots. She is trained in a wide variety of safety procedures, ready to deal with everything from an emergency landing (on the Hudson River maybe) to a passenger-turned-patient having a heart attack. Above all, she knows a great deal about managing people, all kinds of people, in a confined setting that often induces anxiety and restlessness. Managerial skill is a hugely important resource. If you've ever had a lousy boss—and I certainly suspect that you have—you know what I mean. You surely have been in a restaurant or business that was badly managed, and you felt it immediately. Perhaps you have been a bad manager yourself. That flight attendant may never have thought about writing a book called *Management Tips from 35,000 Feet,* but I'll bet she could. And I'd read it.

Then I started to think a little more about Mr. Irate Flyer. Was I too harsh in my judgment of him? He's a complete stranger, after all. Maybe he was having a truly bad day. Maybe he just received some terrible news. Furthermore, I'm sure I've been rude to people at times. I've been tired. I've been angry. This little episode reminded me of how easy it is to blow it, how easy it is to forget that we are all in this together. We are much more alike than we are different. We all have the same humanity and the same rights in the eyes of the Constitution. My dad taught me to treat other people with respect: We should work as hard as we can to view one another with compassion and to recognize the knowledge that others possess.

As the situation diffused and everyone got back to their in-flight activities, I thought more about what I had witnessed. The flight attendant had driven home an important lesson to me about what constitutes a skill—in addition to all the technical information that she needed to have, the safety precautions, and the logistical talent for navigating in such a small space, she needed people skills. Those aren't measurable, they don't really show up on a résumé or even in an interview, but those with people skills have a noticeable and lasting impact on their entire environment. Treating others with respect, patience, and understanding will not only encourage them to treat you well right back, but it also is the first step toward inviting connections, conversations, and collaborations that make everyone's day a little bit better.

Everybody has skills and expertise that are hugely valuable, and we have a responsibility to treat everyone's skills with respect. This means *everybody,* regardless of job title, education level, or social standing. Such open-mindedness fulfills the standards of nerd honesty, and it serves our brute self-interest, as well. But I'm writing this chapter at a moment when a lot of people, in the United States and throughout much of the developed world, are actively resisting the expertise of other people—especially those whom they perceive to be "the elite." There's a terrible misconception these days about the authority that comes with the kind of knowledge that scientists represent.

I regularly hear angry comments about climate scientists and sympathetic politicians. People believe that some of these experts want new regulations so that they can grab more power, apparently trying to dazzle us with big words until they exploit us. In reality, they're just doing their jobs. Some of the most disturbing attacks have been directed against climate researchers like Michael Mann of Pennsylvania State University. Many people have been led into thinking that those scientists want to change the tax structure for their own benefit, rather than doing it so that we can create an energy economy that is sustainable, cleaner, and more efficient than the one we have now. The deniers and conspiracy theorists have lost sight of the basic truth that everybody knows something you don't. In fact, they've weirdly turned the idea on its head. One too-common line of argument goes like this: Who are these "experts" to tell me what is true about climate? They think they know things I don't know—well maybe *I* know more than they do!

Now this is important: Expertise must be earned. Other people know things you don't because they have learned, worked, and lived in ways you haven't. Climate experts are experts because they have spent a lifetime studying, asking questions, and seeking honest answers. I want to go on the record now to make really clear that the actual goals of climate scientists are to make everybody's lives safer and healthier.

Opening yourself to other people's knowledge can be difficult, even unsettling, especially if those other people are very different from you. It can be frustrating to feel like other people know more than you, and it's human nature to feel uncomfortable and vulnerable in the face of confusing information that doesn't make sense to you yet. My angry airplane passenger couldn't see that the flight attendant was trying to take care of him, because he was too immersed in his own assumption that he was being neglected while somebody else got better treatment.

We nerds, scientists, and fellow travelers have a two-part responsibility here. First, we have to fight back against people who actively try to devalue the knowledge that we have fought so hard to gain. I believe we have to defend scientific ideas, and even more importantly, we have to defend the scientific process, the principle of being open to new information. We have to actively promote the philosophy that everyone knows something you don't. That means getting involved with local schools, working on educational projects, speaking openly with friends and family, and engaging with politicians. We also need to have sympathetic discussions (I mean instead of loud arguments) about how we know and believe what we know in science, especially about human-caused climate change. I feel that the "how we know" is a key to moving a discussion along.

Second, we have to embrace that kind of openness ourselves. We have to do more than just pursue the most interesting, nerdiest projects and developments because we think they're cool. We have to think hard about how we apply our pursuit of progress. It can be not just cool but also life-improving and accessible for everyone. I've heard too many climate "debates" that consist of deniers saying climate change is a hoax and the ostensibly pro-science people responding that the deniers are either thoughtless idiots or amoral evildoers. Look, I am very familiar with how frustrating it can be to talk with people who reject scientific evidence, but I'm pretty sure that nobody has ever changed his or her mind as a result of being called an idiot.

We need to defend important institutions like the EPA by explaining exactly what they do and why. We need to hold accountable the leaders of the climate-denialist movement, the enemies of the idea that everyone knows something you don't. We need to expose what they do not know and discredit them. At the same time, we need to find ways to spread information and real evidence in a way that inspires confidence and trust. Wherever possible, we must work to vote the troublemakers out of office, exposing their corruption and offering a clear alternative that will actually protect and uplift us all. And who is this "we" I'm talking about? It is all of us. A meaningful response to climate change will take more scientific research and engineering solutions. It will also take lobbying, public outreach, community organizing, get-out-the-vote drives, and corporate support.

All those actions will benefit hugely from exchanges that, on the surface, may look like hardly anything at all: talking to Karl the janitor about his work or watching a flight attendant soothe a difficult passenger. A true nerd aspires to pull in these kinds of insights all the time.

I'm still working on all this, just as even the best and nerdiest among us still are. I observe human behavior, trying to become a more adept manager at the small businesses I'm involved in. I do my best to hire the best people for each job, but beyond that, I empower the people I work with to do as much managing as they can handle. If someone else is managing a project or job, then I don't have to. That person will soon know more about it than I ever will, and will become a new source of expertise. In the end, trusting and respecting people leads to better teamwork. People get more done and have more fun. When I'm working at The Planetary Society, it means we are more likely to build solar sails and explore new worlds. When I'm working on my Netflix show, it means I am more likely to entertain my viewers, impart some information, and expand their imaginations.

Embracing the idea that everyone knows something you don't is part of getting a great many people working together for the common good. It is yet another part of the path to positive change.

PART II

Nerd Ideas into Nerd Actions

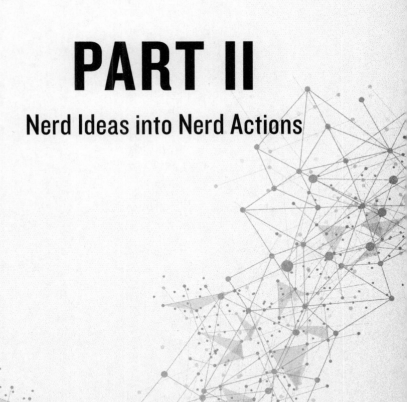

CHAPTER 11
The Joy of Constraints

I'll never forget my first week of high school physics, when Mr. Lang drew an ellipse on the blackboard. An ellipse is a squished or stretched circle; it is, approximately, the shape of an auto racetrack or an especially symmetrical egg. An ellipse is, theoretically exactly, the shape of Earth's yearly path around the Sun. But the shape my teacher drew wasn't readily recognizable. His ellipse was tilted with respect to the X and Y axes, as well as with respect to the floor, the chalk tray, and, well, everything. He singled me out and issued this challenge: Could I make sense of the shape and figure out a way to describe it mathematically? Could I write the equation for a tilted ellipse?

Numbers and equations are used to describe every aspect of the world around us. To do so, though, they need to do more than match the idealized forms of a textbook; they must be able to capture all the complicated, sloppy details found in reality. Equations should make sense without reference to graph paper or coordinate systems. Surely I, as a good student, could figure out how to describe a simple ellipse that had been tilted just a little. So I had at it. Being in a pretty good place with my algebra skills, I thought, well, I'll just add in some tilted terms for the Xs and the Ys, some sines and cosines, maybe. I gave it a go. Well, such an equation can become ponderous for anyone, let alone an 11th grader trying to prove himself at the beginning of the school year. Mr. Lang saw where I was going with this. He stopped me with these words: "Nye, the ellipse isn't tilted; *you're* tilted."

There was a lot of wisdom packed into those few brief words. Mr. Lang was telling me to look at the problem differently. At first glance the problem seemed fantastically difficult, but he was suggesting that the difficulty lay with how I was looking at the problem more than with the problem itself. The secret was to stop thinking about the other objects around the ellipse (the blackboard, the room, myself) as having anything to do with my tilted figure. I could get around the frightening tilted aspect of the drawing on the blackboard simply by looking at it differently. If I reimagined the drawing as existing within a tilted world, it turned back into a perfectly normal ellipse and then I could solve it with ease. Rather than come up with a complicated solution to a complicated-looking problem, I needed to step back and reassess the actual problem at hand. All I had to do was tilt my head, literally.

My shaky algebraic and trigonometric skills prevented me from undertaking the complicated solution I'd initially considered, and that constraint forced me to find what was ultimately a much better solution. We usually talk about constraints as if they are bad things. They prevent us from doing something—often exactly the thing we most want to do. But I'm going to argue here that constraints can be useful, beautiful even. Constraints help guide the decisions in your life, from the smallest to the most profound. Because of those constraints, there are certain approaches to problems that will not work. Constraints help you figure out what to not do and, more to the point, what ideas to leave out. They help you make decisions about the things you buy, the things you eat, the job you take, the person you marry (or don't). They make the world scientifically and mathematically comprehensible. That's what Mr. Lang helped me understand on that seemingly uneventful day.

I talk a lot in this book about the power of considering everything all at once. When it's time to act, though, you can't process *everything*. You can't strain literally every option through the filter of logic; it would drive you mad and take up so much time that you would never end up acting on anything at all. So we have to learn to make decisions with restrictions. We give more weight to some details than to others, and we constantly evaluate what information is most relevant and reliable. Some of the greatest technological triumphs (the Manhattan Project or the *Apollo* Moon landings, for example) took place under severe constraints. That's what we do here in the land of nerds, whether we're facing a small theoretical problem or a giant real-world one—we take the skills and resources available to us, look at all the angles, and do the best with what we have.

One of the marvelous things about the human brain is its ability to sort out incoming information quickly, or quickly enough to keep our species alive through the last few millennia. We can't know everything about the things and happenings around us, so we have to rely on the knowledge at hand and choose a course. Is that a lion on the savanna stalking us for her dinner? We have only a brief time to decide what we might do: run, hide, or *really* run? Programming a robot to carry out a task like this would take a long time, but our brains sort out the sounds, smells, wind direction, and distance to the nearest climbable tree very, very quickly. It's not clear why a robot would be running from lions. Perhaps the robot is actually powered by a delicious battery made of meat.

The same basic skills come into play for soldiers on the battlefield, drivers in traffic, players on the ballfield, or shoppers in the supermarket. In every case, it's about distilling a lot of possibilities into a single action.

A number of studies have been done in which college students were given different rules for submitting a piece of work. Some students got no firm date to complete the task, some got a negotiable completion date, and some got a firm deadline. Consistently, the students who were given the firm finish date did far and away the best. The imposed constraint that they had to finish on a certain date motivated them to budget their time and to focus appropriately on their work. Constraints help direct us to the solution or approach we need to get things done. Without constraints, we tend to lose sight of the important things, just like those unstructured college students did and still do.

There's a great example of the power of constraints unfolding right now in my part-time hometown of New York City. In late October 2012, a pretty good-size storm named Sandy (the still-ferocious remnants of Hurricane Sandy) slammed into the New Jersey shore and parts of New York City, especially the southernmost, lowest-elevation parts of Manhattan. There were blackouts, flooding, significant losses in productivity, and enormous reconstruction costs. The extended shutdown of the economically vital city and surrounding region left an impact on the economy of the whole world. So the states of New Jersey and New York set about looking for engineering solutions that could prevent this kind of storm damage from happening again.

The constraints facing the architects were rigorous. The areas along the coast have to remain pleasant and livable. People have to be able to come and go, commuting, meeting, eating out, strolling waterfront parks, sitting at desks, typewriting books, etc. But when the next great storm hits, all the parks, sidewalks, roads, and subways lines in the areas that are in harm's way have to be durable or flexible enough to get people to safety; then they have to return to being livable and workable again soon after the storm waters recede. Not trivial. Heavy rains and winds cause coastal flooding. Electrical substation switching systems shut down when they're inundated, and subway tunnels filled with water become impassable.

You don't have to know much about rivers and floods to think about possible ways to avoid storm damage. How hard could it be? You build a big, waterproof wall or levee. But here's the thing: If the city of New York were to build a wall high enough to keep out a storm like Sandy, that barrier would have to be at least 3 meters high (about 10 feet). It would also have to be many kilometers, or miles, long. Technically you could do it, probably, but it would look awful and it would not fulfill the livability goal. A long, snaking seawall would isolate the city from the river. We have to rule that out, as a lot of the city's business is on the river. By walling off the river, we'd be cutting off an enormous fraction of the economy. Same problem with blocking the seashore from the businesses of the famed Jersey Shore.

If the concrete contractors were running the government all by themselves, they might say, "That's just how it has to be." The local importers, exporters, waterborne commuters, tour companies, etc., would need to relocate upriver to slightly higher ground. But in our democratic reality, the great-seawall solution had to be dismissed immediately. The nonnegotiable reality that New Yorkers, and people who do business with New Yorkers, absolutely would not accept a continuous seawall affected the whole project. It was a constraint. It made the architects and engineers sharpen their pencils, tap a few more keystrokes, and click and drag a few more computer-aided design lines.

The Denmark-based Bjarke Ingels Group, which got the contract to protect the Manhattan waterfront from future storms, went to all kinds of trouble in their design to satisfy all the practical and aesthetic demands. They created a plan for a series of contiguous designs to address a potential flood, as it would affect each neighborhood from downtown to uptown. On a map oriented with north at the top, the designers envisioned a 10-kilometer-long (6-mile) series of parks, tunnels, berms, and roadways elevated and reinforced to withstand the great wash of the floodwater that will accompany the inevitable next superstorm, as well as the ongoing rising of sea levels. Shaped like a horseshoe, the overall design has come to be called "the Big U."

Along with the constraints of having to keep the area suitable for business, maintain or improve livability, and make use of existing highways and streets, the architects also had a lot of information to guide them. The data on damage caused by Superstorm Sandy are detailed and stark. These engineers have a pretty good idea of what can go wrong. The rebuilt waterfront neighborhoods have been designed to function normally almost all the time, but when a big storm arrives, the revised shoreline must absorb the floods. The next giant storm should leave the area largely intact and completely livable. There will still be plenty to clean up—when an urban flood subsides, it leaves behind a lot of trash and other unpleasantness—but city planners believe the neighborhoods and subway trains will be in much better shape than they were after Sandy.

Money is another constraint, of course. The cost of the Manhattan waterfront-rebuilding project is estimated at $335 million, according to Bjarke Ingels. That may seem like a lot of coin, but it's about $\frac{1}{200}$th of what Sandy cost the city in lost business and damaged infrastructure. The Big U will be well worth it, if it works as planned.

» » »

Most of the time we have no difficulty identifying the problem we want to solve. The hard part is figuring out why the problem is a problem at all. Defining that cause then clarifies the constraint and makes the whole thing easier to solve. In the case of my ellipse, the cause was that I didn't understand rotated coordinate systems. If you look and see that your basement is flooded, for instance, there are many different possible underlying causes. Is it that the drainage is inadequate and you need to install a pump? Is it that the previous owners failed to seal the walls on the east side of your house? Or is it that climate change is bringing on too much rain every year, and this is no longer a sustainable place to live?

This is where outside expertise—the nerd collective—becomes indispensable. In my math class, I didn't have to look far; Mr. Lang played the role of the informed outsider. In the case of my hypothetical basement flood, I would assess the cause as well as I could, but then I'd probably call on people with complementary knowledge. I might even need to go through a sequence of expertise, from plumber to building engineer to environmental scientist, before I had adequately defined the cause and isolated my helpful constraint. Calling on the experts does not reduce your control of the situation; it increases it. I feel like a lot of people misunderstand this point when they complain about the "experts" telling them what to do. Experts help you constrain the problem and move forward. Without them, you can still do something—but there's a very good chance it will not be the right thing.

When I'm not in New York, I'm generally in Los Angeles working at my job as the CEO of The Planetary Society. The expertise around me there—just wow. My colleagues had to convince me I could do the job, and sometimes I still can't believe I'm the guy in charge (see Chapter 21 for more on that). The Planetary Society is the largest public organization supporting and advocating for space exploration. As the head guy, I do my best to guide space policies all over Earth toward the goal of finding evidence of life on another world. I feel strongly that such a discovery would be among the most profound events in human history. But to most legislators, regardless of nation, sending robots to drill for microbes on Mars sounds like a distinct luxury—and a costly one at that. Reconciling these two perspectives is a case study in another type of constraint.

Sometimes the biggest constraint we face is how to get attention for the problems that we want to solve, especially when our solutions require time, money, effort, and/or attention. There was a time when geopolitics alone was enough to drive such exploration. The *Apollo* voyages to the Moon were made for the sake of winning the Cold War; they never would have happened without the US and the USSR competing for superiority. Those of us who believe deeply in the value of space exploration have to sell it in new ways to overcome the modern political constraints. We have to think rigorously about what's important in space and about the benefits here on Earth.

A big part of succeeding as a nerd is figuring out how to tell a story about a problem in a way that gets others excited about being part of the solution. For instance, I strongly believe that space exploration brings out the best in humankind: We achieve mighty things when we venture above the atmosphere and send our best instruments, designed by our best scientists and engineers, to make discoveries around and upon other worlds. We are at our greatest when we push against the greatest constraints.

There's education, for one thing. Space exploration is a potent motivator that helps draw kids into science and technology. The US Department of Education spends nearly $80 billion every year. For comparison, NASA's planetary science budget is about $1.5 billion— about 1/50th as much. Ask yourself: Which one excites more young students and inspires them to tackle the hard work of calculus and Advanced Placement physics and chemistry?

There's also technology. Here I'm talking not just about the spinoff inventions from NASA and NASA-related programs, although they are numerous, from fuel cells to digital cameras. Consider the much broader value of the World Wide Web, weather forecasting, and the global positioning system. But far, far more significant for me is that space exploration is how we learn more about ourselves and our place in space. By exploring worlds beyond our own, we solve problems that have never been solved before. And when we get used to solving problems that have never been solved before, the world becomes a better place. Space exploration has created a culture of innovation that affects everyone on Earth every day.

You might notice that there's a feedback loop of constraints at work here. Dealing with the technological constraints of launching space probes to other planets forces engineers to be extremely creative; that creativity, in turn, helps address the political constraints that often prevent such missions from getting funded in the first place. All along the way, enlisting nerds to work on problems of space exploration yields all kinds of secondary benefits to society. It strengthens the entire chain of expertise. Very crafty. Very everything-all-at-once.

The battle against constraints can seem exciting and inspirational when we're talking about high-minded space exploration, but it's a different story when the conversation turns to practical concerns. Even a focused project like New York's Big U took a great deal of unglamorous negotiating and cajoling. When it comes to bigger challenges, constraints tend to lead people to despair, pessimism, and inaction. It's all too easy to fixate on the paths that are cut off from us and think that there is no way forward at all. From there, it's a short journey to denial, or to a nostalgic yearning for an earlier time when our problems seemed less difficult.

If you know me at all—and by this point in the book, I'd bet you do—you can probably guess that when I think about progress in one area, I'm thinking about how it applies in other areas and disciplines, as well. And I'm always thinking here about our rapidly warming world and our rapidly changing climate. A large number of people today believe, or say they believe, that climate change is not really happening. There are politicians and business leaders who claim we'd be better off if we tracked backward toward the age of coal and oil. They find the constraints of our energy future too scary to contemplate. But I am confident that this crisis of confidence will pass. We have overcome many similar crises in the past.

Space exploration forces scientists and engineers to expand their thinking because of the extreme nature of those constraints. For protecting Manhattan's waterfront from future storms, the constraints were set by practical considerations such as cargo transport and pedestrian access. For the kinds of missions championed and funded by The Planetary Society, the constraints are utterly impractical, sometimes comically so. A project might begin with a brainstorming session along these lines: "Okay, everyone, you have to land a car on Mars. Let's go!" This is no made-up example. It's the very problem that NASA had to deal with in creating the *Curiosity* rover that is currently rolling across Mars, and will deal with again at the end of the decade when the agency sends an even more advanced rover, currently called *Mars 2020.* Both rovers are about the size and mass of a Chevrolet Spark automobile. So how are they going to do it?

If you have the naïve confidence of a budding engineer, you might think, "It can't be all that hard. We just have to slow down enough to roll or skid to a stop. We land airplanes all over the place every day. We landed all sorts of things on the Moon. Surely we've got the basics of that figured out by now." In other words, you'd start with the problem you know, just like I started out trying to solve the ellipse using the math I'd already learned. But it turns out that this business of setting down intact on the surface of Mars is some kinda crazy complicated. On Earth you have a lot of air to work with, and even the fastest fighter jets are dealing with much, much lower speeds. When the probe carrying the *Curiosity* rover approached Mars, it was moving at more than six times as fast as an F-35, with the throttle to the firewall—going all-out. That's a lot of energy to dissipate.

Over the years, the creative engineers at NASA designed a few lovely retrorocket systems. Just as in an old *Flash Gordon* serial, the 1970s *Viking* landers lowered themselves tail-first onto the surface of Mars, retrorockets flaming beneath them (I mean, dude, those rockets were, like totally *retro*—uh, sorry . . .). As cool as they are, retrorockets are not an affordable option for gently landing a rover, however. Rockets kick up a lot of dust, which could damage sensitive instruments and moving parts. They also blow out a small crater where the blast hits the ground; getting out of that crater could present problems for the rover just as it's getting started. And all that fuel and the gizmos to steer the exhaust are heavy. So much for familiar solution #1.

On to #2, using the atmosphere to slow you down. The two *Viking* landers had heat shields that dragged through the atmosphere, scrubbing off some of their deep-space velocity so the retrorockets didn't have to work as hard. If a heat shield can slow things down a little, then a parachute should slow things down a lot, right? Ah, but this is where the in-betweenness of Mars messes with you again. It turns out that the Martian atmosphere is much too thin for regular Earth-style wings and parachutes. The atmospheric pressure on Mars is 0.7 percent of the pressure here. There just aren't that many air molecules for a parachute to catch hold of. Furthermore, an incoming spacecraft and its parachute will be going faster than the speed of sound in the upper Martian atmosphere. Entering at supersonic speeds sets up pressure changes and shock waves that can make a parachute rip itself apart. It would be a sonic boom of death.

So there was no existing piece of technology that would accomplish the job at hand. The engineers at NASA's Jet Propulsion Lab in Pasadena had to tilt their heads and look at the problem in a different way. The constraints of the unprecedented conditions and existing technology forced them to create something that had never been created before, mixing the old with something completely new. They identified the problem, zeroed in on the causes of the problem, blended expertise, and achieved beautiful progress. They drew on ideas not just from earlier space missions but also from military jet research and from automobile safety testing. You can see why the constraints of space travel are so constructive.

A breakthrough came when the engineers realized that they didn't have to get the spacecraft to a complete standstill; all they had to do was make sure the landing was gentle enough to survive. They started with a "cap and ring slot" supersonic parachute to slow things down a bit, followed by a retrorocket phase to slow things some more, and finally a landing phase aided by . . . four big balloons (talk about old technology). No kidding. Right before hitting the ground, the capsules holding the Mars rovers puffed up with four super-tough airbags. They freely fell the last several meters and bounced more than a dozen times along the equivalent of a few football fields over the red dusty surface before deflating and opening up so that the rover tucked inside could drive away. NASA put three rovers, *Sojourner*, *Spirit*, and *Opportunity*, on Mars that way, with spectacular success.

The fourth rover, *Curiosity*, was considerably bigger and more capable than the ones sent before. With more weight came a new set of constraints. Airbags would not be enough. Now the engineers had to tilt their heads again, and they emerged from their brainstorming session with an even kookier solution. Their insight this time was that retrorockets are okay as long as they don't get close enough to the ground to kick up big clouds of Mars dust. Their solution: Start out by slowing way down using a specially vented supersonic parachute. Then, fire retrorockets, but don't let the exhaust nozzles get too close to the surface. Next, lower *Curiosity* from the underside of an eight-rocket jetpack, which the engineers call the "Sky Crane." Have the Sky Crane fire its retrorockets, bringing the whole package to a near-standstill, suspended about 20 meters (60 feet) above the Martian surface for a few seconds. Then the one-ton rover could rapidly rappel down three nylon ropes, like an Army Ranger descending from a battlefield helicopter. Once the rover is safely on the ground, the ropes would detach, and the Sky Crane would blast off again, flying away to a safe distance before crash landing. Hard to believe, but this system worked beautifully, too.

While we're on Mars—no, wait, we're still on Earth. I mean while we're on the subject of being on Mars, we must contemplate what we would need to do to put humans there. For a crewed mission, landing a one-ton payload will not be nearly enough. We'll be trying to deliver dozens of tons, maybe 30 or 40 tons, at a time. We will need to send enough equipment to build a full life-support system and backup systems, complete with protection from radiation and from the harsh Mars environment. We will need to send food and medical supplies. All this on top of the scientific equipment needed to study the planet and to search for Mars microbes (Marsrobes?), whether living or fossil. We'd have to look at those constraints and conclude that sending astronauts makes sense even after we take them all into consideration.

That's a whole new set of constraints to overcome, by far the most difficult ones yet. It's like the earlier question about landing a car on Mars but a whole lot more so. Right now, there is no conclusive answer.

Engineers at NASA and at SpaceX, the company run by Elon Musk of Tesla fame, are still tilting their heads at the problem, but they have a promising concept. Since parachutes and retrorockets both have limitations, they thought, how about a system that combines the best aspects of both? The system they have in mind would shoot retrorockets at just the right angle and at just the right velocity to act like a gigantic virtual parachute in the thin Martian air. Testing this concept will be hugely challenging. High-altitude rocket experiments on Earth are the best nearby approximation we've got. Ultimately, it will take a real landing on Mars to prove it would work. If they succeed, it will be the result of the same lessons I absorbed back in high school. Knowing which approaches can't work makes it easier to find the ones that can. If they fail, that will feed new constraints into the next effort.

I bring this all up to make a point about the process of creating progress. Each of the designs I've discussed here—airbags, the Sky Crane, the retrorocket parachute—adds to the tool kit for space exploration. Each constraint inspires a new solution to draw on for the future. I am confident that someday we will land probes on other intriguing worlds, including Jupiter's ocean moon Europa and Saturn's moon Titan, which is dotted with lakes of liquid methane and ethane. When that happens, engineers will make use of that tool kit. If none of the existing solutions are quite right, they'll just tilt their heads again. And every bit of the increasing collective expertise will help in overcoming constraints on Earth, as well.

Less than a century ago, people worried that smallpox was unbeatable and would eventually kill every human on Earth. In the 1790s, and again in my lifetime in the 1960s, many people thought that population growth would overwhelm our ability to produce enough food, leading to widespread starvation. Strange as it might seem now, many people thought that the year 2000 computer-clock problem (Y2K) would bring our society to a standstill—and maybe it would have if engineers had not hustled and addressed the problem in every computing machine they could find. It was not a magic spell that kept Y2K from being big trouble; it was diligence and attention to detail on a big scale.

Climate change is a bigger challenge than any single problem we've faced before. I hope it inspires more creativity and more dedication than ever before, becoming part of the nature of progress. As our ambitions become greater, so do the constraints we must deal with. Then those constraints drive us to tilt our heads and find new, bigger, more creative solutions. That's true for each of us as individuals, and that's true for society as a whole. But we cannot afford to stop, or even to slow down. We need to meet our constraints as they confront us.

As soon as the going gets tough, I am confident that tough people will get going. Soon, other coastal cities will have to start rebuilding their waterfronts just as New York City is doing. That work is starting to happen as a reaction to devastating natural disasters that are landing all over the world. As soon as the first floors in Norfolk, Pensacola, Galveston, and Miami are ankle-deep with a few centimeters of water day and night, people will take all this seriously. They will be able to draw on engineering work pioneered in New York and other early-reacting cities. But just think about what kind of damage and tragedy we could prevent if we started solving the problems before they happen. From a scientific and engineering point of view, we can see them coming clearly, and we have more than enough information and evidence to anticipate the constraints that will need to be addressed.

Instead of running in circles, waving our arms—or, worse, going about our business in willful ignorance—we could get to work now. We could erect wind turbines off the east coast of the United States, Canada, and Mexico. We could install photovoltaic panels practically everywhere the Sun shines. We could heat and cool a lot of our dwellings, offices, and factories using geothermal sources. We'd create jobs, boost the economy, clean the air, and address climate change. If you really want to make America great (and the rest of the world, too), these are the main things you, I mean we, need to do. It sounds like an enormous undertaking, and it is, but as we've seen again and again, the enormous ones begin with small perceptual shifts.

I'm sure you have encountered tasks that seemed to be impossibly complicated—until you tilted your head, took a moment to think, and got the job done. We all have. Most of the time it is something small but instructive, like trying to figure out that vexing chalk drawing of an ellipse on the blackboard. Every once in a while, though, we encounter a truly great challenge. Then those smaller ones offer inspiration about how to proceed. They show us how constraints can make problems seem less overwhelming and can help guide us toward the most workable solutions. That's the situation the whole world finds itself in right now. We need to draw on those crucial life experiences. Instead of despairing or evading, we need to embrace constraints, face our problems, and get to work.

CHAPTER 12

Upside-Down Pyramid of Design

Young people often ask me what advice I would give them to help them succeed. My answer reflects lessons I learned in my early days working in engineering. If you have any nerdish proclivities whatsoever and like to tinker—which, at some level, practically everyone does—I encourage you to get in at the beginning of a project. "Be part of the start" is one of my beloved aphorisms. When you join a conversation at the inception of an idea, whether it's solving an existing problem, starting a new company, or designing a new product, you can influence the development process and know that you did everything you could to be part of something good. Of course, it's a risk. You might be part of something not so good, or something that simply sucks. But it's better to take that risk and do your best than it is to work on a mediocre something or other that leaves you feeling like an old Ford Pinto or Chevy Vega. (They were crummy cars.) But I'm getting ahead of myself.

At this point, I was working at Sundstrand Data Control with a very sharp, if somewhat curmudgeonly, designer named Jack Morrow. He was amazing. He was in his fifties, which seemed to me quite old back then; now I realize, of course, that it is actually the very prime of youth. Despite his "advanced" years, Jack was a helluva skier, quite an athlete. His mind never stopped moving, and neither did he; as he talked, he continually removed and replaced his reading glasses. While the managers kept themselves busy shuffling memos and moving around the boxes on their company organization charts, Jack was doing more than thinking about how to work through constraints. He was mapping out a whole organizational scheme of his own.

Jack must have sized me up as a fellow who could use a little sage advice, so he talked with me often about the importance of getting the design right. "If the design is bad," he'd say, "no matter how well everyone else does their job, the result is never going to be any good" (or at least "never as good as it could be"). At Sundstrand, I was on a team that was designing drilling equipment, heavy stuff, and Jack strongly encouraged me—no, he pretty much insisted—that we make sure that everything really fit before we asked the machinists to cut metal. He, ah, drilled the point in my head: Work it through and be absolutely certain it works correctly on paper before you commit someone else's time, someone else's skill, and someone else's materials to a finished part or assembly.

What Jack understood is that it is easy enough to get things *almost* right. Even for designers and engineers working on big-budget, large-scale machinery, there's a constant temptation to stop at the point when things are 90 percent good. Laziness is part of the reason, but not all of it. When you find a workable solution, even if it is only somewhat workable, you begin to experience a mix of relief and satisfaction at another constraint overcome. You know what I'm talking about: It's like the feeling you get the moment when the lawn is almost raked, the dishes are almost all washed, the second coat of paint looks okay without any final touch-up . . . It takes a lot more effort to get things exactly right. But as I learned working alongside Jack at Sundstrand, things that *almost* work do not exactly work, which is to say that they do not work at all. I left a few months before the company got sued for declaring nonworking avionics boxes to be real working boxes in its salable inventory—a form of cheating on your taxes. Not everyone there thought like Jack.

Whenever you set out to create something, be it a hook to hang your coat or a string of computer code that lands an airplane, you have to take time with the design. The more time you can set aside to think, Jack reminded me, the better the created thing will be. These are two crucial skills for everything-all-at-once progress: Filter information carefully so you can home in on the best ways to solve your problem, and then develop your ideas fully in the hypothetical before you execute, so that the resulting system really does what you intend it to do.

Easy enough for me to write, but it's often quite difficult to accomplish.

So there I was, young Bill, laboring away at Sundstrand, creating devices to help direct mining and drilling operations. Meanwhile, I was also absorbing what Jack and my colleagues discussed regarding the general state of domestic engineering in the United States. At the time, particularly in the automotive industry, it seemed as though we were falling further and further behind our international competition. This was a cause for great concern for Jack and the others.

I was not, and am not, an automotive designer. I am a mechanical engineer who loves knowing as much as I can about mechanisms. But for nerdy engineering fun, I routinely dropped the transmission, pulled the engine, or replaced the constant-velocity joints in a friend's car or truck. Car culture was much more of a thing back then. Even my parents were drawn into it. Because of our French ancestry, we ended up with a few Renaults in our driveway, including the Renault 16 in which my brother and I installed our first THANKS sign. Even as a high school student, I could see that these little French cars were full of clever innovations. The materials might have fallen short, but the ideas were cool. They had front-wheel drive, which frees up more space for passengers because there is no need to make room for a driveshaft connecting the engine to the rear wheels. They had disc brakes for better stopping. Rack-and-pinion steering. MacPherson struts, a compact and efficient suspension system. A rear compartment made bigger by mounting the two rear axles or half-shafts side by side rather than one above the other. From around 1970 on, I began to realize that Detroit was not keeping up with what designers overseas were doing. The Europeans and the Japanese were innovating more than we were.

All through engineering school at Cornell, I felt frustration with what the US automotive industry was doing—or, more accurately, not doing. Right about this time, American automakers created those two famously terrible small cars, the Ford Pinto and the Chevrolet Vega, that continually reaffirmed my low opinion of them. To do routine maintenance on the Vega, you had to loosen the motor mounts so that you could tip the engine to get at one of the spark plugs. The engine block was made out of aluminum; the head, or top, plate of the engine was iron. That was General Motors' attempt at progressive engineering, but the aluminum block tended to warp at high temperatures, causing it to leak lubricant and coolant. It burned oil, ground itself to a standstill sometimes, and generally made a mess of things. And the fender rust—you could count the days to when it would surely appear.

The Pinto was even worse, notorious for bursting into flame when hit from behind. Managers at Ford knew about the risk but convinced themselves that reconfiguring the gas tank so that the car was less vulnerable to rear impacts was not worth it. The Ford managers were morally wrong, and they turned out to be economically wrong, as well. At least 27 people were killed in fuel-tank fires, at least 117 lawsuits were filed, and Ford's reputation was greatly diminished for years.

In the decade that followed, imported cars, especially those from Japan, drastically increased their share of the US market. It's no coincidence. Detroit started with bad designs and then let things get worse from there. My colleagues and I talked a lot about this troubling turn of events. Jack had been my age during the *Apollo* era. He had developed mechanisms and systems that helped land people on the Moon. And here he was watching the whole country roll downhill, leastways in domestic engineering. He wanted to do his part to push back and encourage good design, and with that in mind, he shared with me a very memorable piece of wisdom.

Jack made a sketch, which I've done my best to live by over the years. He called it an "upside-down pyramid," a triangle with one vertex pointing to the bottom of the page. I embellished his sobriquet into the "upside-down pyramid of design," for obvious reasons, and refined it into a master plan for setting yourself up for design success. We divided this pyramid into horizontal layers, each of which represented a step in the design or production of a thing. The upside-down pyramid of design is not (yet) a nerd icon like the slide rule, but it should be. It is a visual shorthand that illustrates the best way to turn ideas into action.

Since I've been going on and on about cars just now, we'll start with one of those as our example. Down at the bottom vertex is where the design takes place. Automotive designers take into account a great many things: drivetrain, number of seats, safety features, and overall look. How will the handling and interior space change if the engine drives the rear wheels instead of the front ones, for instance? How many fasteners, how much paint, and how many tires or hinges will you need? It's all got to be figured out. In general, design is where the fewest number of people are involved. This is the cheapest step in the process of bringing a product to market. It's just people sitting and thinking thoughts about requirements, shapes, materials, looks, and feels. But once that is done, then your car company (or construction firm, or movie studio, or whatever entity is involved) starts spending real money, which is what makes this phase the most important one, as well.

Above the design on the upside-down pyramid comes procurement: getting the stuff you need to make something. For a car, there are sheets of steel, plastic, glass, rubber, wires, and so on. In an architectural undertaking, it would be cement, wallboard, insulation, and glass. For a seamstress or seamster, it's when you buy the fabric, cut it from the bolt, and start shearing shapes that you really commit to a pattern or design. That's when there's no turning back, from a financial standpoint or a sewing-table stool. This is the level where you start to really shell out the cash. Someone has to get all those things. We're talking about the purchasing department and manufacturing engineers, people who figure out where to get the materials, which have to be of a certain quality, in a certain quantity, at a manageable price, available in the right place in a manageable amount of time.

In the case of the design, you're paying salaries to the designers. But on the procurement level, you're paying salaries *and* you're buying stuff. If it's a car that you plan to produce millions of, there's millions and millions of dollars' worth of raw car stuff. There are important

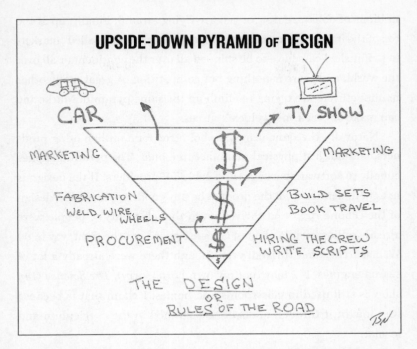

decisions to be made at the procurement stage, as well. The materials you choose have a huge influence on your final design. Cheap cloth can ruin the cut of a well-tailored suit. Beautiful new buildings can be marred by ugly mineral runoff from cheap prefab brick cladding. Hard, shiny plastic can make the interior of a luxury car feel like the low-rent option from the airport parking garage.

Then you go up to the next layer of the pyramid. You start using the raw materials to make parts: wheels, fenders, bumpers, transmissions, and gas tanks. Go up to the next layer, and you'll find welders and painters. Move up again, and you start fitting all the purchased parts together. Now we're really spending money, because all these steps require skilled laborers paying attention to what they're doing. Millions and millions of dollars, euros, or yen are flying around. Finally, up at the top of the inverted pyramid is the mythic layer broadly called "marketing." Finished cars have to be shipped all over the continent or all over the world, and there is nothing but competition. A great many other manufacturers are trying to climb up the same pyramid. Marketing can make or break your vehicle's success.

Naturally, the same principles hold true for a million other products, and not just physical manufactured ones. The pyramid applies equally to software packages and cable TV providers. If the design is not good, neither shall the product be any good. Apple got the design of the iPhone right and thrived, even though plenty of people were already selling mobile phones. Google figured out the right way to do Web searches and thrived, even though there were already a lot of search engines. If I may toot my own horn (sorry), *The Science Guy* show is still used in classrooms and homes. I claim that is because we thought it through and were disciplined in the curriculum and format.

In each case, there is the amazing thing about that pyramid: The whole wide tottering top—the place where a creation finally sees the light of day and captures the imagination of the public, or not—depends on the little triangle of design at the bottom. You could have your car painters come to work and do calisthenics together to be an ace painting team. The welders could come in to work and sing the "What a Great Day to Weld a Quarter Panel" song, crooning like Sinatra, taking tremendous pride in their jobs. The upholsterers could lovingly install the finest reasonably priced seats money can buy. The wiring team could lay those wire bundles in with graceful curves and cute little harness tie straps. But if the initial design sucks, the best thing that all those people will come together to create, even on your very best day, is a crappy Ford Pinto. You're never going to get anything better from this assembly line than what was designed on the white board, graph paper, or design tables.

If the design is no good, no matter how hard everyone works, the product will be no good, either. On a television show, you can have the charmingest host ever, with the greatest director ever, with the quickest, sharpest camera operators in the business, but if the idea for the show sucks, the show is going to suck. The videotape storerooms and hard-drive servers of TV networks are fully stocked with shows that never made it past a few episodes because the initial design or conception didn't get done well.

On the other hand, if the design is great, then you have a solid shot at an equally great product that people will take pride in creating and that consumers will want to use or watch. That was the indelible message I took away from my experiences at Sundstrand. A good design doesn't guarantee a great product, because there are plenty of places to go wrong in execution; but you will never, ever have a great product without a very good design.

There's another saying, one I picked up while I sat on the Seattle Bicycle Advisory Board: "Good engineering invites right use." For many years, I've commuted in big cities by bicycle. It's so much more pleasant than sitting in a car or on a train, so long as everyone understands how the bike lane is supposed to be used. But if a stranger is driving along a street with a bike lane, will he or she be able to instantly figure out where the bikes should ride and where the cars should drive? A lot of dangerous or deadly bicycle–car interactions happen because of avoidable confusion about lane markings; drivers should always know exactly where they need to keep a lane clear for cyclists, and there should always be unambiguous signs or lights indicating who has the right of way. The examples of inviting right use are infinite. The instruments in an airplane cockpit should be self-explanatory. I've often mused that when I come upon a door handle or bar, I should never have to guess whether to push or pull. Embrace the idea, and you have a shot at creating something of real value.

As hopeful as each of us might be about the beginning of a project, remember that things generally don't come out right the first time through. Even with this upside-down pyramid design idea, you are almost certainly going to run into glitches with your project, be it a dress pattern, a new spacecraft atmospheric braking system, or a software module. But when you're in the design phase, that's okay! It's all too easy to get swept away by someone else's sales pitch or to go along with the will of a large group because it's the path of least resistance. To that I say: Don't do it. Accepting other people's values without question is exactly what gave the world the Pinto. Before you start spending money, apply the core nerd values instead. Ask yourself: Do you honestly believe in the course of action? Does it address a meaningful problem? Does it invite right use? Would it make the world a better place? If it succeeds, will you be proud of it? If it fails, will you have learned something in the process and be glad that you tried?

It's why I strongly encourage everyone to be willing to take a second shot, to have the expectation that even if the first thing you build is good, it's probably not going to be good enough to sell to a stranger. In my opinion (once again correct, obviously), you almost always have to build the second prototype.

Spend time with the initial design, but budget and plan to spend more time with your second bite at the apple. Trust that you will make mistakes, and plan on learning from them. It's a hard lesson for most of us to learn because it costs the one thing so few of us have—time—but it will save you much more time (plus energy, money, and reputation) if you commit to making every conceivable improvement before you move to the next phase of development.

Taking that time to evaluate your work and find ways to strengthen it makes all the difference. A small vehicle that was created around the same time as the Vega and Pinto was the Datsun B210, produced by the company now known as Nissan. It was made of the same raw stuff as its American counterparts, but it worked better and ran well over 100,000 miles. The designers started with a similar concept (a small, economical car), but they just did more with what they had. Because of their quirky looks and extraordinary reliability, those cars won a devoted following. Then, more than a decade after the B210, Mazda, another Japanese company, came out with the Miata. Like the Datsun, the Miata resembled its competitors in design and materials but trounced them in execution. It matched the looks and handling of the British roadsters that inspired it. The big difference was that it ran flawlessly all the time. Now in its fourth generation, the Miata is the bestselling sports car in history.

What I perceived as a decline in US engineering and manufacturing skill had a deep effect on me. I watched America's automobiles fall behind those produced by the rest of the world. This was around the same time that America's shipyards, steel mills, and many other heavy industries ran into trouble, as well. I watched it happen from the inside, working at a high-tech engineering company that had directed me to make avionics boxes that did not work and could not be sold. I was disenchanted, and I wanted to do something different. I loved engineering, and I especially loved my native country, but I was very concerned about the future of both.

I've thought a lot about what went wrong. The United States is a huge country where companies could continue to find buyers for a long time even if they sold substandard products. In many cases, those companies had dominated their industries, leading to a false sense of superiority; engineers and managers alike couldn't imagine that competitors were equaling and then exceeding their own products. Meanwhile, management and labor often fell into adversarial relationships. Many factors, with one overall outcome: The failure to acknowledge mistakes, continually pursue new standards of excellence, and reject inadequate designs led to steady losses and the eventual closing of the doors. I thought a lot, too, about the future of the United States. I realized that really changing things will take years and years, but it can be done. If we focus on young people, someday the United States will be back on track, doing great engineering work and making great products that could change things.

When I left Sundstrand, I took all that hard thinking about good design and did something with it that, on the surface, made no sense at all. I committed to doing stand-up comedy. I thought seriously about what I wanted to do and what I wanted to contribute to the world. I'm not saying I was funny, but I was trying. I kept going down a new path that led to *The Science Guy* show, to *Bill Nye Saves the World*, to this book. To my way of thinking, designing a system or gizmo that someone thinks is valuable is exactly like writing a bit or constructing a show format that someone thinks is funny. It's in the bones or the scaffold or the chassis.

A television show might not seem much like an automobile, but the principles of good design apply everywhere. The only way a show with a host is going to work is if that show is an extension of the person on the marquee. *The Tonight Show Starring Jimmy Fallon* is an extension of the playful nature of Jimmy Fallon. *The Late Show with Stephen Colbert* is an extension of Stephen Colbert's irony and wit. Same thing with Chelsea Handler's *Chelsea* and, I hope, *Bill Nye Saves the World*. My show(s) have to be an extension of me and my view of the world in order for them to succeed. They have to be true to the design at the bottom of the inverted pyramid. With a clearly outlined vision for our goals, our structure, our tone, and our message, everyone working on the project is on the same page and moving forward with a shared vision.

Blending comedy with science remains my passion. Your passion is surely something that reflects your distinctive knowledge and experiences and desires. Your job may look very different. Still, we're all working with the same rulebook. Great things can happen if you have a clear idea of what you're trying to design and if you are part of the start. Good engineering invites right use.

Right now, we have a great many impulses—in politics, business, engineering, and really every area—that push us toward cheap fixes, short-term solutions, and rushed decisions. When things go wrong (and they usually do under those circumstances), often the result is a lot of useless finger-pointing. It's easy to get discouraged. But as I have learned over the years, there is also profound joy in seeing your concept all the way through, past the 90 percent mark, as far as you can take it within practical limitations. It's the nerdy thrill of rejecting mediocrity. Using the upside-down pyramid as your guide, you can stay focused on the things you need to do to make the world a better place . . . by design.

CHAPTER 13
Comedy and Me

If I could somehow show you newsreels of my childhood, you'd have no trouble seeing how my sense of humor (if I indeed have one) came to be. My father was funny. So was my mother. My sister laughs like no one else I've ever seen. She gets to the point where she can't breathe she's laughing so hard. I often report that my brother is the funniest man I know; I mean, he's funny to laugh with rather than simply funny-looking. See? Comedy is that simple. But as anyone who has tried to make an audience laugh on cue can tell you, comedy is also quite complicated. It requires empathy and perspective. Being able to find humor by stepping outside your normal point of view is an important skill, and not just for making your sister gasp for air. It is, I'd argue, a vastly underappreciated tool for shifting the way you look at life's problems and for finding novel solutions.

I was fortunate that humor was valued in my family to the point where it seemed completely normal to spend our days trying to outdo each other at being punny. Jokes and comedic replies were part of how we went about our days. If someone said, "I'm going to take a shower," someone else would crack, "Be sure to put it back." If you don't grasp that joke, irony is not your strength. Maybe it's more aluminum-y . . . uh, sorry. There it goes again. The silly, never-ending, why-don't-you-stop-saying-that-stupid-shower-joke thing continues to this day. I do it automatically. When I was a sophomore in college, my roommate Dave Adams got tired of "Be sure to put it back" and retorted, "No, I'm letting it go right down the drain." My family uses his reply even now, decades later. Which tells you that my family is still making the same shower joke decades later, too.

As a kid, it took me a while to appreciate that humor and comedy do not get nurtured in every family. As I got older, I discovered that people who become my friends are, as a rule, pretty funny—it may be more of a guideline than a rule. But I also noticed that some people are serious all the time. (I have a feeling you have no idea who you are.) Furthermore, some people, more than others, can produce or induce a good laugh. (We know who *you* are.) I wondered, Why the difference? Eventually I realized that there isn't much difference after all. People everywhere enjoy a good laugh, regardless of their comedic ability, inclination, or background.

Humor is especially prevalent among the nerds, I'm pleased to say. There's a distinctly nerdy quality to comics from Groucho Marx to Steve Martin to Louis C.K. Why are so many nerds funny—or maybe I'm asking, why are so many funny people nerdy? The popularity among the nerd set of one particular style of humor—punning—offers some clues. You may have noticed that the irony–aluminum-y joke is the same kind of humor I described early on in this book, when I shared my deep abiding love of the Greek letter phi (φ) and all the science-themed "phi fi fo fum" jokes that come with it. At the time, I was mainly thinking about my φ phixation as an expression of the connectedness of knowledge. Now I'm realizing that there's something else cool going on there, as well. Punning, by its very nature, connects thoughts that you wouldn't normally put together. (If you didn't connect *iron* and *irony*, then you missed out on that particularly hilariously funny joke and . . . well, I'm sorry for you.) The resulting clash is the thing that makes the pun funny, but it is also a serious creative act. It trains the brain to adopt a conceptual flexibility, separating possible interpretations from the literal presentation of a word or an idea. Piecing together distantly related thoughts in new and unexpected ways is one of the most important sources of creativity, both in the sciences and in the arts. At least it should be, *oui*?

What I'm saying here roundaboutly is that punning is a form of information play. Of course nerds are drawn to it! Even as they joke around they are making connections, and as with any skill, the more they do it the better they get at it. There's a funny positive feedback at work: Humor encourages new connections, and new connections open up the mind to more kinds of humor. It's a nerdy process, though it certainly does not belong exclusively to any group. You can engage in this kind of mental exercise by opening your mind to the rich and absurd juxtapositions all around you. You may be surprised by the new connections you make. You might even make somebody laugh (even if that somebody is just yourself).

The perspective-shifting power of humor also makes it a hugely effective coping mechanism in the face of adversity. I'm sure you've experienced what is often called "gallows humor"—the relieving laughter that erupts in tense or sad situations. It is like the punning process, but it plays out a different way; it reinterprets events until it finds the kind of absurdity that triggers a laugh. That is definitely where some of the Nye family humor came from. On December 8, 1941, Pearl Harbor Day (west of the international date line), my dad and his comrades were attacked by the Japanese Navy. On December 24, 1941, my dad and his company were captured by the Japanese Navy on a remote atoll called Wake Island in the middle of Pacific Ocean nowhere. He spent 44 months in a Japanese prisoner-of-war camp. I've been to a couple reunions of the surviving Defenders of Wake Island, and after listening to their stories, I concluded that it was my father's sense of humor that got him through.

Being a prisoner of war was, uh . . . stressful. Every day these guys were subjected to beatings. Every day they were hungry. Every day they were exhausted. In summer, they worked in oppressive heat. In winter, they were chilled to the bone. Early on, a sailor was picked pretty much at random and beheaded with a sword in a weird reenactment of a 17th-century Edo ceremony, just to show the prisoners that their captors meant business. My father described the guards in the camp as "gung ho." Yikes! That's it? They cut off a guy's head and it's just "gung ho" (a strongly Americanized version of a Japanese term meaning "work together," which became a catchphrase in the US Marine Corps)? But that was how my dad and his guys handled things—things that were too hard to talk about openly at the time and that remain so today, even with their wives and kids.

Playing with language became an important way for my father and the other POWs to keep their spirits up. They cooked up an entire fake language called "Tut" to prevent their Japanese captors from understanding their private conversations; they became information play-guys. In the Tut language, you spell words out letter by letter very quickly: If it's a consonant, you pronounce it "[consonant]-u-[consonant]," so the letter "b" becomes "bub," and "f" becomes "fuf." You say the vowels in normal fashion: "a, e, i, o, u." Certain letters are exceptions: "c" is pronounced "cash," to distinguish it from "k," or "kuk." After 4 years of this, everybody was pretty fast with Tutting it up. My dad's buddies would pronounce his name, Ned: "Nun-E-Dud." Hey, Nun-E-Dud, wow hash e roy e i shush tut hash a tut shush hash o vuv e lul? ("Hey, Ned, where is that shovel?")

Years later, when my father taught us the Tut language, it was merely a silly-sounding word game that we played together to see who could speak the fastest. But at the time, under the guard of the Japanese, such word games could get quite serious. My dad and the other veterans of that camp were reluctant to describe or even acknowledge their war experiences, so I can't be certain what really went down. But because Ned Nye and his buddy Charlie Varney were so fast with their Tut language, I figure it was more than just a pastime for them. I suspect that the prisoners alerted each other of dangers, like an approaching guard, by making succinct warnings that sounded unintelligible to their captors. I would not be surprised if that Tut talk saved some of their lives. It certainly helped keep them sane and focused. Turning bad into good is part of what makes humor so precious.

During the starving, gut-wrenching, teeth-clenching, white-knuckling reality of day-to-day life in the prison camp, there was the US Marine captain who was nominally in charge of my father's contingent of guys. By my father's account, this captain was pretty insecure, which he covered up with swagger and what he thought of as an imposing-sounding vocabulary. In particular, he was given to peppering his sentences with the term "disirregardless." Now people—this is not Standard English, and we're not even talking about the common nonstandard "irregardless." The captain added an extra "dis" up front, crafting a word of true nonsense value (although it is a triple negative that mathematically circles back to a real meaning). To my father, who was quite the man of letters, hearing his captain huffing out this nonexistent word was a hardship that, over time, grew into an irritant that rivaled the hunger, battering, and all the rest of it. It became the focal point for a sanity-saving perspective shift, one that allowed him to focus on a small absurdity instead of an enormous terror.

Channeling my father's sentiment as best I can, his reaction went something like: "This war sucks. Prison camp sucks. It all sucks. But now . . . now, I have this guy and his 'disirregardless;' this *really* sucks." Somehow this word became among the funniest things in my father's life. It let him step outside his whole awful situation, enabling him to find some relief, an essential distraction that connected him to his wordsmith identity that existed and mattered to him outside of and in spite of his captivity. To be able to take offense at "disirregardless" in the face of such horrific circumstances allowed my father to stay connected to an essential part of himself, the civilized, articulate, human part of his identity that had existed back home and that would, he hoped, soon reassert itself once the war was over.

Everybody was stuck together in the same situation—prisoners in the POW camp, not knowing what was happening to the rest of the world during the war—and there was no point in rehashing the US Navy's missteps or lost opportunities that landed them all there. That would have been too real, and pretty much not funny at all. Instead, my father found humor in his helplessness and went after "disirregardless." The release of tension is one of the fundamental drivers of humor, and the captain's blustering became another long-running family joke. My brother, sister, and I still say "disirregardless," often shortening it to just "dis-irr" [diss-ear], to convey something akin to the currently popular "I'm just sayin'."

Which brings me to the third perspective-shifting aspect of humor. In addition to changing the way you look at ideas and situations, it changes the way you look at people. Many classic forms of comedy force you to stand outside of yourself and to think about how others see the world, or force others to look at the world through your eyes. It's the essence of ironic detachment: A priest, a rabbi, and a monk walk into a bar. The bartender looks the three of them up and down carefully and says, "What is this—a joke?" It's funny because it's familiar but still manages to take you on a sudden left turn. (In fact, that's a second-level joke, which forces the listener to adopt the point of view of somebody telling the joke to a listener who already knows an earlier version of the joke. Heavy stuff.) I assume my father could imagine the captain's thought process, his absurd insistence on throwing around important-sounding words in a futile effort to feel like he was still in control of a totally out-of-control situation. More important, my father could help his fellow POWs see that absurdity, as well. It let them all share a laugh and, in the process, help maintain the tight social bond that was essential to survival.

These are some of the key lessons I absorbed growing up: Humor is a playful way to experiment with novel ideas. Humor is a way to displace anger and stress. Humor is a way of forging a deep connection with other people. Humor can redeem some of our darkest experiences.

During my childhood, I wasn't consciously aware of these things. I just knew I enjoyed being funny and liked the way people responded to it. When I was a high school senior, an English teacher approached me about playing a part in our school's presentation of *The Taming of the Shrew*. I scored some good laughs as Tranio and developed my fondness for stage comedy. Meanwhile, my brother, Darby, introduced me to the remarkable art of the comedic monologue. Darby was fascinated with Johnny Carson's opening monologue and ability to talk about everyday things in a funny way. My brother and I got in the habit of watching Johnny's monologue on Friday nights before going to bed. I paid close attention to see how Johnny did it—one man entertaining his audience all across the country.

Many years later, my longtime friend and colleague Ross Shafer told the story of having coffee with Johnny Carson. Ross asked Johnny how he (Johnny) had managed to stay on television for 30 years. Johnny said, "I never tried to be the best guest on my own show." Instead, he took a genuine interest in his guests and kept a thoughtful distance from himself. He had that golden currency of comedy: perspective. He empathized with his guests, and just as important, he empathized with his viewers, too, inviting them to feel as if they were on the show with him. There is an intense vulnerability and honesty that comes with this approach, since it takes away the protection of hiding behind a false persona. When Johnny did comedy, he brought the perspective there, as well; that's what made his monologues so compelling to me. His humor was the bonding kind, not the divisive kind. It's the same with Louis C.K. There's a certain type of comedian who brings the audience into the circle with them to create a strong sense of camaraderie. They let the audience laugh *with* as well as *at* them.

When I went to college, science and engineering mingled with comedy and performing in my budding, wrinkling brain. I partnered with Audrey Moreland, a civil engineer (a nerd like me, in other words), and we entered the Cornell talent show, doing a jitterbug routine that we made up. We finished fourth, but man, the applause! It was huge. We took our show on the road—well, to a bar next to campus—and tried out a local version of *The Gong Show*, which was quite popular for a few years. Audrey and I did okay. We won a couple of dance contests, etc. One day around that time, my friend Dave Laks came hurrying over to my house babbling, "You've got to see this. You've got to see this guy." (Dave and I had been roommates as freshmen. He went into materials science while I went into mechanical engineering. I know what you're thinking, "How exciting . . ." Well actually, it was in its way. Dave is yet another funny nerd.)

It turned out that Dave and his housemates had this brand-new technology called "cable television." Cable began to change the way information influences our culture. With broadcast television, there were basically three national channels in the United States at the time— ABC, CBS, and NBC, plus some odds and ends. Cable began to break that up into many more streams, which meant that each one could be more specialized, giving people in the United States and around the world access to all kinds of videos and performances they never could have seen before. It was peanuts (small) compared with the kind of data proliferation that happened in the 1990s with the World Wide Web (large), but at the time, cable was a revelation. And what Dave wanted me to see was something unfamiliar that might not have ever made it to one of the three basic channels: a video of Steve Martin at the Boarding House nightclub in San Francisco.

It's not just that Steve Martin was funny. He was funny in a different way than we had seen on TV before. And he was funny in a way that seemed oddly . . . familiar. As I watched, Dave was saying, "Look at this! Look at this guy! He's just like you!" I thought Steve Martin's perspective on our world was brilliant. I wanted to be part of that picture. I felt that I shared his same sense of irony or absurdity. And yes, I confess: I confidently thought I had his same sense of timing . . . or tie-ming. Whether I ever did is up to you, the audience, to judge. Steve Martin's absurdist bit about expecting that his audience was full of plumbers who'd laugh madly at punch lines involving "The Kinsley Manual" and a Langstrom 7-inch wrench was oh-so-close to my hilariously funny bits about 316 stainless steel. Wasn't it? Well, I thought so.

I realized that comedy in public didn't have to be limited to Johnny Carson addressing the whole nation. Every bit of Steve Martin's comedy was carefully packed into his seemingly off-the-cuff monologues. There was one bit that really made an impression on me because it cut right to the heart of comedy—that anxious awareness that some people in our society are on the inside and that some are on the outside. Steve Martin launched into a riff in his mock-sincere style. "Remember when the world blew up, and we all came to this planet on that giant space ark? Remember, the government decided not to tell all the stupid people, because they were afraid that . . . " And there is that little half-a-beat before the crowd in the nightclub gets the joke and starts to laugh. It was brilliant. We are all, the whole audience, the stupid people on the outside. Check it out: an enormous shift of your perspective in $12\frac{1}{2}$ seconds. Not bad.

I see a common thread in all these different examples of humor. They all play with insider and outsider viewpoints to create a shared experience. There's not a cruelty of mocking one perpetual outsider, because the comedian leads the audience between both sides—sometimes we're all the outsider, and sometimes we're all on the inside. This movement back and forth, in the hands of a skilled comedian, can open the audience's eyes to different perspectives and shake them up a little bit. Steve Martin remains a genius at shifting between mock-arrogance and mock-humility, carrying his audience along with him. My dad was a powerless prisoner in a POW camp, but at least he and his buddies were in the inside group that knew what the word "regardless" meant. Being able to move fluidly between insider and outsider allows you to look at the world from a lot of different angles at once, which enables you to see a whole lot more.

◈ ◈ ◈

From the moment my friend Dave first showed me that Steve Martin video, the engineering-versus-comedy tug-of-war inside me began. It kept going all through my early career. But because science and engineering are wonderful, too, I still loved my day job. I loved to watch the dimensions add up to create perfectly mating parts (oh, come on, people—I'm talking about engineering plans) and to feel a drawing take shape with my hands. Working on a long drawing board, designing those wonderful mechanisms, is a specialized kind of craft, beautiful and elegant in its physics and precision. It was 1978; I was a young guy; I wanted to do some good in the world. I was fascinated with the guys who could visualize and design the elegant linkages, the mechanisms that connect the cockpit controls to the movable control surfaces of the plane. Did you know that you can fly a 737 airplane even if both engines quit? The linkages enable the pilots to use the energy in the moving air to move the control surfaces, which in turn steer the plane. How could you not want to be around the kinds of people who think of things like that?

Engineering and comedy, two of the great nerd skills, fought for dominance of my time and attention. Steve Martin's success was amazing. His albums went platinum. One became so popular that Warner Bros. Records sponsored a Steve Martin look-alike contest. My new Seattle friends were thoroughly exposed to my obsession . . . er, my "abiding interest" in the Wild and Crazy Guy. I guess those pals liked what they heard, because they pressured me into entering the contest. I went to the venerable Peaches Records & Tapes to sign up (in the original vinyl-record era). A short time later, I drove to the long-ago-burned-down Montana's nightclub, where the contestants each performed some stand-up in Steve's (I call him Steve) style. I did it, and I won. Next up, the gentle folks at Warner Bros. flew me south to San Francisco, where I competed against guys with a great deal more stage experience—and I was eliminated. Nevertheless, I was hooked. I am pretty sure I wouldn't be here writing to you as Bill Nye the Science Guy if not for my one peculiar connection to the brilliance of Steve Martin.

After my modest success as an impersonator, various promoters asked me to do a Steve Martin routine at parties or corporate gatherings. I did a few, and I was okay—but just okay. I wanted to do my own comedy, write my own jokes, and get my own laughs. I tried and tried. I went to open-mic nights at comedy clubs, when anyone in the audience can get up to try out material, and gave it a shot. Now and then, I would get a real laugh, and it was addictive. Prodded by the rapid growth of cable TV, every major city in the United States and Canada soon had at least one or two stand-up comedy clubs. The standard format is a three-person show: the emcee, who warms up the crowd, the middle comedy act, often called just the "middle," and the headliner. I dreamed of middling. I would get home from my day job (working on a drawing board, technical document, or computer keyboard) and immediately take a nap. Then I'd wake up around 7:30 p.m. and head downtown to perform at open-mic shows.

At those open mics, I first met Ross Shafer, a former combination stereo–pet-store owner who ended up hosting several network shows, along with a second comic named John Keister. Together, those two ended up changing my life. About 2 years later, we started working together. At one of the local TV stations, KING-TV, there was a program director named Bob Jones; he was the guy in charge of what show gets put on when. It happened that Bob hired Ross to host a show called *Almost Live!*, with John cast as a version of an independent correspondent. John then asked me to play a crazy person in a sketch in which Ross got kidnapped. I was funny enough, apparently, as a nut job wearing a pyramid-shaped hat made of bendy-straws. I was developing my own sense of the absurd.

For whatever reason, Ross and John kept me around long enough that I eventually got up the nerve to quit my aerospace job. It was October 3, 1986, roughly. No wait, that's exactly when it was. I had

$5,000 in the bank. I figured that was enough to sustain me for 6 months even if I didn't make any money at all in the interim. I had a modest mortgage and so on. It felt like a risk I needed to take. I think my friends regarded my decision as an interesting choice, but they tried their best to be supportive. When I was not being especially funny doing my stand-up set (as your time on stage is called), it was hard for them. That happened . . . uh . . . more than once. To a person, they encouraged me to keep pursuing my passion, although they also advised me to keep my hands on the board. By that I mean keep engineering in my back pocket in case I needed it. So I did.

My biggest fear at the time was that I might fail and be both unfit as a comic and useless as an engineer. It was an imposter-syndrome moment for me (more about that in Chapter 21). *Almost Live!* was broadcast for 26 very nonconsecutive weeks a year. I continually worried about the other 26 weeks that had to be filled with something that made money. I was fortunate that I could get work as a contract engineer. I treated my life as a design problem and did my best to make the parts fit together. I did freelance drafting work at a couple small engineering companies in the Seattle area between writing bits and jokes for *Almost Live!*, doing stand-up in the area clubs, and appearing on camera.

At this point in the story, Ross Shafer was not only the host of *Almost Live!*; he was also the host of the most popular evening drive-time radio show in the Seattle area. This was so long ago ("How long ago was it?"). It was so long ago that the hottest station in town was still on AM radio. For his show, Ross would write fake interviews with fake people and then perform all the voices himself. I listened every day, often from my drawing board at my freelance job working for Avtech, a Seattle aerospace company that makes cockpit displays and airplane dashboards. I designed a liquid-intrusion-resistant radio knob for airplane cockpits—a fancy way of saying you could spill coffee or a soft drink on them and they'd be okay. Basically, I applied my engineering skills to whatever task I could get them to send my way.

Well, one day while I was listening, someone (an actual listener!) called in to Ross's show to answer a question about the first *Back to the Future* movie. Something about the amount of electrical power needed to send the DeLorean traveling through time. The correct answer, Ross explained, was 1.21 "jigg-uh-watts." Well, I couldn't leave that uninformed utterance unchecked. I called in seconds later to explain that, in science, we prefer to say, "1.21 gigg-uh-watts," with a hard "g." It was silly, but attention to language runs in the Nye family, and every good nerd knows you need to get your terminology right. It turns out that the listeners (Ross, at least) found my awkward precision entertaining. One thing led to another, and I ended with an assignment to call in to Ross's show every day at 4:35 p.m. to give a vaguely science-based answer to listener questions.

A little idea started to incubate in the recesses of my mind. I wasn't consciously thinking it, at least not yet, but it was there taking shape: The connective power of humor is a great way to get people to pay some attention to science. A lot of folks will tell you that science is boring or alienating or just no fun, but here I was getting people to listen and laugh. I was playing with ideas, and a bunch of radio listeners— many of whom probably never thought of themselves as nerds—were digging it. Furthermore, I felt the United States was losing its edge in science and technology. Maybe I could help, just a little.

Then came the next link in my great chain of serendipity, or something like that. In January of 1987, I was attending a writers' meeting of *Almost Live!* when a guest canceled. As I remember it, the absent guest was Rita Jenrette. (She was notorious for claiming to have had sex on the steps of the US Capitol Building; even weirder, it was sex with her own husband.) When Ross tells this story to his audiences, he says the guest who vanished on us was Geraldo Rivera. When the very funny producer of *Almost Live!*, Bill Stainton, tells the story, he says it was Eddie Vedder, lead singer of Pearl Jam. It was a busy time; memories blur. In any case, we needed to fill 6 or 7 minutes of show time. In the world of television, that's a lot of airtime. Imagine staring at a blank screen for as long as it takes to boil two eggs—consecutively. Ross and the rest of us were somewhat desperately casting around for ideas. ("How somewhat desperate was Ross?") He was so desperate that he said to me, "Why don't you do some of that science stuff you're always talking about. You could be, I dunno, 'Bill Nye the Science Guy' or something." In an instant, my two separate nerd lives were smashed together. No, let's say my lives "merged."

My first idea was to do some kind of sketch with liquid nitrogen, which is great for making fog and shattering flowers. Working with the head writer Jim Sharp, we came up with "The Household Uses of Liquid Nitrogen." I thought it was funny. More amazing, *other* people found it funny. I won a local Emmy. After that, I was tasked with coming up with Science Guy bits about every 3 weeks. I realized that I could make impressive effects happen on TV if I was willing to fake it, essentially performing magic tricks. Magic is a reliable crowd-pleaser, and it's a skill well suited to nerds who pay attention to details that others overlook. I tried making it look like a grapefruit could generate enough energy to run an electric motor, which is possible only with the help of hidden wires. But the tricks and magic were not satisfactory to me—not at all. I wanted to show people real things they didn't realize were possible, and I wanted the constraint of doing real science. Comedy and science have a lot in common: They both depend on the perspective shift that comes with real-deal honesty.

The success of the *Almost Live!* segments slowly gave me credibility as a performer and, to a lesser extent, as a writer. From there, one thing led to another. Two KING-TV employees, Jim McKenna and Erren Gottlieb, started their own production company and hired me to host *Fabulous Wetlands*, an educational video for the Washington State Department of Ecology. I guess they picked up on my many references to environmental stewardship, riding a bike to work, etc. More jokes were written, more time passed, and then we created the pilot for the *Bill Nye the Science Guy* show in 1992.

You might know a little about how this part of the story turned out. I poured my heart into it. I loved science; I loved humor. Being able to connect with an audience using comedy and science to make them appreciate both a little bit more—that is just the best. And the response to our show continues to amaze and thrill me. I am always curious to hear what it is that people say they enjoy most about it. I've had many people tell me over the years how much they liked it as "educational television." Well, *The Science Guy* certainly was and is intended to be educational. But keep in mind, the show was absolutely, every time, always created as entertainment first. If a television show is not entertaining, viewers are long gone. A surefire way to make something entertaining is to make it funny—though that works, of course, only if you *can* make it funny, like actually funny.

For *The Science Guy* show, I'd do almost any silly, old-fashioned gag to get a smile from the audience—or since they were the ones who were actually right in front of me as I performed, from the crew. During a single-camera shoot, such as *The Science Guy* show or the rehearsals for my new *Bill Nye Saves the World*, the crew is usually the only audience around. It's especially satisfying when I can trigger a "crew smirk," a real laugh that emerges as the crew try to suppress a laugh of any sort. I've managed it even at Fox News, where they try not to laugh at anything. Goaded by my crew, I reached deep for laughs.

Pie in the face? Oh yes. Slide headfirst into real, heavy wooden bowling pins? Sure. Get stuck in mud and make jokes about not being able to get out . . . because I really couldn't? I guess so. Buckets of water thrown at me? That one's pretty obvious. Those slapstick segments were coupled with more sophisticated, dignified moments such as when I tripped over cables and fell on my face while making a serious point, or screamed in fright when I saw a human skull that I pretended to forget that I was holding in my own hand. Science offers a window into humor, but sometimes humor offers an unexpected window into science and human nature. Whatever the gag was, if the crew laughed, we knew we were onto something.

A practical note: If you are ever going to get a bucket of water thrown at you while you're wearing a lab coat (as would be standard procedure), untuck your shirt first. An untucked, starched cotton shirt will deflect quite a bit of the incoming water around your belt. Your pants, and specifically your crotch, won't get nearly as wet as they would, or do, with your shirttail tucked in. This became a small, unexpected tutorial in fluid dynamics.

My antics may seem far removed from the wordplay of my father and his Tut language, but I maintain that the two are not that different. They both rely on the quintessential nerdy sense of self-awareness and on the connective power of the resulting perspective shifts. What kind of man would happily utter a phrase like "Wow hash e roy e i shush tut hash a tut shush hash o vuv e lul?" What kind of man would allow himself to be abused over and over in the name of science? Someone who is willing to drop (some of) his ego for the sake of the audience, who is willing to get under a joke for a little while in order to impart a larger lesson or sense of community.

Part of the laugh that comes from watching a performer doing slapstick is a kind of internal relief: a little voice in your head that says,

"Hold it; didn't he see that coming? I'm glad that's not me, 'cause I'd never do that. Wait—would I?" For a moment, you are outside of yourself in a more aware and awake altered state. At that moment, the guy wearing the lab coat is not a remote authority figure. He's a sympathetic guy, unifying the audience around him, allowing them (even without knowing it, perhaps) to find great joy in Newton's third law. And once that guy wins your sympathies, you become more receptive to what he has to say and more interested in the things that interest him. Humor and sympathy, it turns out, are excellent tools for winning someone over to your point of view. I still rely on them every day.

The journey from being Ned and Jacquie Nye's little kid to being Bill Nye the Science Guy looks like the result of a lot of coincidences, but I see an organizing principle here. First, I've come to believe that there are no "big breaks;" there are just breaks. I followed paths as they appeared and did my best to keep my preconceived notions and pride from getting in the way of my curiosity. I worked to avoid blinding myself to what might be around the next corner. It seems like I've used every scrap of knowledge I had at my disposal to make my way through each situation, regardless of whether or not it seemed like a natural fit. I became an engineer because I liked bicycles and bumblebees, and it looked like a route to a career with a steady income. I started making people laugh, I think, because it was a natural part of the shared language of nerds, and I got on stage so that I could make more people laugh. I tried stand-up comedy as a result of having made several other people laugh, and then I found my calling as a science educator in a comedy writers' meeting because it let me share my two passions equally. Every step of the way, I worked on the craft in front of me while doing my best to keep all the other pieces of the puzzle nearby. Look at your own successes and I'll bet you see the same kind of methodical procedure at work.

My journey as a comic is far from complete. High on my wish list is hosting *Saturday Night Live*, where Steve Martin worked up some of his greatest bits. I especially loved "Theodoric of York, Medieval Barber," in which Mr. Martin's character would start by recommending antiscience remedies like bloodletting for his patients. Then he would pause and launch into a what-if vision of enlightenment, one that would sweep away all his brutal practices and replace them with a description of our modern scientific method. He'd reach a crescendo of excitement—and then dismiss it all with a derisive "Nahhhhh." Theodoric expressed the two sides of what is inside all of us: an impulse toward greatness with a heavy yoke of laziness and self-doubt. In seeing the absurdity of this character, it is impossible not to wonder what we all might achieve if we made a radical embrace of reason. It makes us laugh at our reasons for holding back.

Comedy and engineering were my path, but in many ways, my story can be your story, too. I hope you enjoy punning and playing with new words. Try to see the world from other people's perspectives. Find the comforting humor in your setbacks. Think about how a joke can bring a group of disparate people together. In short, be a comedy nerd. Using comedy to break away from old preconceptions opens you up to innumerable insights you would miss otherwise. It is exactly what you need to improve your design skills, whether the thing you are designing is a 747 control surface, a job application, a watercolor painting, or a great sight gag. It opens you up to connections and collaborations with people you might otherwise overlook. It's a gateway to expertise. Oh, and it can make life a lot more fun.

Once you break through and laugh, you are liberated. You also tend to become well liked (or well-enough liked) by those around you. Leastways, that's what I've heard. Very nerdy can be very funny. What's not to love?

CHAPTER 14

Not Faking It

What is the primary industry of Hollywood? Blockbuster action movies? Fashion and glamour? Celebrity gossip? I maintain that the real industry there is storytelling. Everything else, including the enormous flow of money, begins with a well-told story. That makes Hollywood a fascinating and instructive paradox. Almost all those stories are fictional (even the "based on a true story" stories often involve a great deal of imagination), and yet they also have to be rigorously honest. Engaging and holding a viewer's attention is no simple task. If a story does not feel urgently, compellingly, rivetingly *true*, the audience goes away.

In case you opened straight to this page, in 1990, 2 years before launching *The Science Guy*, I took my plunge into that great paradox. I was working 26 weeks a year on the *Almost Live!* comedy show while still freelancing as a mechanical engineer. I wasn't sure what was next for me, but I was open to new adventures. Those are exactly the times when adventure happens. Steve Wilson, the director of *Almost Live!*, had a friend named John Ludin, who was a producer in LA. John, along with the writer Bob Gale, was creating a *Back to the Future* Saturday-morning cartoon show for CBS based on the movie trilogy that had just concluded. The concept was that every episode would have an educational component. The characters would get into a predicament, and then they'd have to resolve their troubles with, you guessed it: science. Having seen my bits on *Almost Live!* (and liking them—go figure), John asked me to be the on-camera guy for short, live-action educational videos that would run in the middle of the animated main show. Fact and fiction, side by side.

Now I should tell you, for me, there's hardly a better, more fun science fiction movie series than *Back to the Future*. (Hear me out, *Trek* and *Star Wars* fans.) First of all, the three interlocking stories all involve time travel and far-out technology, and who doesn't love all that? Second, the heroes of the series are creative thinkers—cunning everykid Marty McFly and his time guide, the wild-eyed inventor Doc Brown—while the villains are the kind of people who care only about money and power. Third, *Back to the Future* presumes an optimistic world of the future, with ever-improving science and enough human ingenuity to figure out how to harness science for good rather than for evil. When John contacted me, I said yes in a heartbeat.

Suddenly I had a regular gig on national television as the presenter in what was called the "Video Encyclopedia" in the *Back to the Future* cartoon. It was a fine idea. The producers hired Christopher Lloyd, the actor who originated the role of Doc Brown in the movies. He has a marvelous, absentminded-professor manner and voice. With Marty and his team stuck somewhere in time, Doc Brown would suggest we look this or that up in the Video Encyclopedia. Then the show would cut to a very noncartoon Bill Nye in a white lab coat, gray shirt, and signature bow tie. Without speaking, I'd demonstrate how to make batteries out of lemons or why a curveball curves. In those days, camera operators would shake their heads at pure-white clothing, believing that it was too bright and overwhelmed the video fidelity in any scene. So I, as the Science Guy, wear a light-blue lab coat (always have). When producers and costumers hear "lab coat," they typically go for a white one, unaware that there's a whole light-blue world out there. I shake my head.

In those *Back to the Future* vignettes, generally no more than 2 minutes long, I came up with the demonstrations, so I had the chance to control things. I thought, "This is cool. I can be an educator on TV." One of the most important things in science is nerd honesty. If you fudge your research data, that is not merely an act of fraud. Falsified data disrupts the entire process of testing hypotheses and finding better answers to questions about how things work—whether those things are distant stars, brain tumors, or the hydraulic linkages on a 737 airplane. You put a lot at risk by letting results that are even a little bit wrong move forward by accident, and you should never *ever* deliberately let bad data be treated as fact. So I automatically assumed that my science education Video Encyclopedias would be honest, too.

In one of the early animated adventures, our heroes had an encounter with shocking high-voltage static electricity. As the on-camera demonstrator, I approached this scene like a standard science educator. I hung a balloon on a string from a small laboratory stand, just a vertical metal rod and a clamp to let the string-slung balloon swing freely in response to a positive or negative electric charge. This is a classic demonstration. The idea is to rub a second balloon on your hair, in this case my hair, and build up a static charge; then I would bring the charged-up balloon near the hanging balloon. When this experiment is done normally, the balloons carry enough electromagnetic charge to repel each other at first but then attract each other after the hanging one spins halfway around. The big finish is to pull one balloon with another one. The generally negatively charged balloon in your hand attracts the complementarily positively charged one. An observer (like the viewer at home) can easily see the movement. Whether I am in a classroom or the Irving Plaza theater, I then explain what was and wasn't happening in some hilarious way.

We were all set on the set, about to shoot the compelling and amazing, albeit pretty small, movement of the balloon due to static electricity. But before the shoot began, the director decided that he wanted to reposition one of the shades on one of the lights to highlight the effect. Meanwhile, I stood there and waited with charged-up balloon in hand and standard hair stylists' "product" in hair. When the director finally called, "Action," not much happened. I moved the balloon in my hand toward the one hanging by the string, but there was hardly any interaction. The problem was that we had waited too long. While the camera was close up on the balloons, the director hadn't wanted me to move at all, leaving me and my balloon hanging. Air, especially moist air, is slightly conductive, so static charges typically dissipate after a few minutes. By the time the director was done making the endless lighting adjustments that unchecked gaffers will engage in, the balloon in my hand had lost most of its charge to the surrounding air.

Now, I'll admit that in this demo the balloon often doesn't get pulled very far, but there's always some swinging and spinning—and it's real. For whatever reason, though, the assistant director was in some kinda don't-pay-attention-to-anything-else-we-gotta-move hurry. He immediately called for "the glue." A stagehand showed up and sprayed adhesive from a can onto the hanging balloon. After that, as you might imagine, the balloons stuck together like crazy. Under pressure to perform, and being the lowest link in this chain of command, I went ahead and smooshed one balloon into the other to get them to stick together, with the glue doing all the work. But it wasn't real. They were not sticking to each other because of the electrical attraction from static charge, which was the whole point of my demonstration. Instead, I was basically miming what the experiment should have looked like if done correctly. It was a small thing, but I still regret letting it happen. To this day, if I watch that bit, I just shake my head. It does not look right, not at all. What in the world were we doing there if not to capture the science, the real action, and show the viewer what we were describing?

You might say, as most of the people on the set that day said, "Bill, get over it. It's a small thing." My, oh my, I did and I still do worry about it. Science depends on honesty, but what I had presented was basically a lie. Even a little lie is still a lie. I turned that negative experience into a positive, you might say. I vowed then and there that if I ever got a show of my own, we would show only real science. To this day, I use this story to remind the people I work with now that the science has to come first. Fortunately, I did get a shot at (re)running an experiment with static electricity. It's in *Bill Nye the Science Guy* episode number 25.

By showing only real science on the kids' show, we established a trust between us and our audience. Like magic, much of our video entertainment these days features tricks, which are almost fibs. People fly on electronically erased ropes. People run at "vampire speed" on winch-driven plastic belts through the woods. Films and television exaggerate so much that viewers pretty much assume they're seeing an exaggerated view, which creates an expectation of deceit. I think that cynical expectation has contributed to the widespread cultural attitude right now that almost every perception or experience is subjective.

But that ain't how it is, in science especially. Dedication to reality—to showing the amazing things around us in the real world—is my way of fighting back. It's my little statement that there are such things as inviolable truth and facts and that objective reality is well worth pursuing because it's exciting, enduring, and essential for creating a healthy and safe world.

On *The Science Guy* show, keeping it real sometimes took extra effort. That static-electricity demonstration wore me out; it was surprising how tired I was at the end of the day, if you get my drift. For the show, I was exposing myself to a Van de Graaff generator: a rubber-belted, static-electricity machine named for one of the guys who helped develop the atomic bomb. You may have seen one at a science museum, where a demonstrator asks for a volunteer with long hair to put her (or his) hand on the device. With the volunteer standing on an insulated platform, her hair will stand dramatically on end, each fiber repelling the adjacent ones. That's exactly the effect I wanted to demonstrate, so I donned a long wig—the "Rocker Wig of Science," I called it—with the intention of making my hairs stand on end in crazy-man fashion. I am confident that everyone reading along here has at some point in life received a static-electricity shock. You may have reached for a doorknob on a dry winter day. You may have put on an especially sparky sweater. You even may have gotten a pretty good shock from the hot wire of an electric fence for a garden or horse pasture. I'm sorry for any discomfort I may have dragged up by reminding you of the occasion. In any case, it's one thing to get a shock. It's quite another thing to get hundreds of shocks during the course of a long day of video recording, take after take.

But after all the work and discomfort, we got the experiment exactly right. We showed the balloons attracting each other. We showed how a lightning rod works. We did not fake any part of it. Was setting up every static-electricity shot for real more trouble than spraying glue on things? In the pragmatic sense, yes. In the meaningful, authentic sense, no, absolutely not. I strongly believe this gave the show an authenticity that viewers admire and appreciate. Certainly it was and continues to be very much appreciated by my beloved colleagues in science education.

In analogous ways, we all have to define our place of authenticity and always be prepared to defend it. You will meet people who want things to happen quickly and are willing to cut corners to do it. Just as many people try to race through the bottom part of the inverted pyramid of design, they often settle for compromise in the overall honesty and truthfulness of the project or product. And by "they," I include myself. You, too, I'd bet.

When it happens, I urge you to embrace the honesty and integrity of the nerd culture. In the short term, faking can often be faster and seem simpler, but it is never better. You may run into problems halfway through and have to start your project all over again. Worse, you may get all the way to the end and find that you have produced a Ford Pinto—something far more serious and dangerous than a dishonest demonstration of static electricity, but born from the same desire to get results fast and keep things moving forward. Either way, you are short-changing yourself in the process. The authentic approach forces you to understand your thing or idea fully and to explore the optimal solutions to any problems you encounter. It encourages innovation, and it helps you become aware of any fakery that might be around you.

Note that authenticity is integral to the way science works, at least when it is working right. The great power of the scientific method is that it demands evidence to back up an idea. You can think of it as the nerd code of conduct. Scientific frauds might sneak by for a while, but in the long run, the truth always wins out. Probably the most famous hoax in history was the 1912 "discovery" of Piltdown Man, an alleged bridge species between humans and apes. For a period, it was an accepted species, although many scholars were dubious from the start, and years of sleuthing conclusively proved that the remains were a fraud. A more recent and pernicious example of fraud came from Andrew Wakefield, the British medical researcher who published a paper claiming a link between vaccines and autism. Other scientists became suspicious when they couldn't reproduce his results; later investigations revealed that Wakefield had altered his data. His paper was retracted and he was barred from practicing medicine in the United Kingdom, but his false claims continue to bolster anti-vaccine suspicions. The more you embrace authenticity, the easier you will find it to sniff out the frauds, whether they are frivolous or potentially fatal.

The benefits of rigorous nerd honesty are hardly unique to science. Think about your friends who are open and direct with you. Aren't they the ones you feel closest to? Think about the people on your crew or in your office who have instant credibility when they speak. Aren't they the ones you want to work with, and isn't that because you trust them to do the right thing? Following the nerd code of conduct makes life easier and more pleasant all around. Not that nerds are immune to mendacity and duplicity (if only it were so), but authenticity land is hostile territory for those kinds of things. I do my best to stay in the zone, and I find it a daily source of relief. As a design principle, it is so simple: Don't say or show things if you know they aren't real! I'll admit to exaggerating sometimes, especially if I really want my exaggerations to be true. But I do my best to keep myself grounded in reality, and I remind myself often that what's most amazing is when you can work within the constraints of reality to achieve amazing results.

» » »

While we were shooting season 1 of *Bill Nye the Science Guy* for the Disney Channel, we fell a few days behind in getting an episode shipped to the Disney technicians in Burbank, California. The company dispatched two well-suited executives to visit our studio in Seattle and find out why. They walked onto our set, which was in a warehouse near the waterfront, and immediately thought they saw the problem. They said, with barely veiled condescension, "Why did you guys 'run brick' all the way up there?"

It took me a beat or two to comprehend their concern. These big-time TV people were used to Hollywood soundstage sets that could be rapidly disassembled and reused. They were looking at the Nye Labs set, which was two-plus stories high, and couldn't understand why we'd wasted so much time and money building an oversize fake brick wall. I eventually managed to explain to them that we hadn't "run brick" on a knockdown set; we were shooting in an actual building made of actual bricks. Well, my Disney execs were a little embarrassed, but I didn't blame them for their mistake; we got along well enough once I cleared up the confusion. They were so used to the artifice of those fake sets, so accustomed to a particular way of doing things that included fake brick and spray glue, that they weren't able to recognize a real brick wall when one was in front of them. Really, almost any one of us could make a mistake like that. We live surrounded by little glowing screens displaying flashy, escapist entertainment. Even sensible people start to think they understand combat because they've watched movies about World War II, or they embrace dubious stories about an inventor who created a warp drive because something similar happened on TV. Fakery can blind us to the world right in front of us, leaving us unable to see the brick wall—until we smack right into it.

So in *Science Guy* land, we kept merrily going the opposite way, fixating on every scientific and engineering aspect of the show. One day, we were doing the episode about digestion. (It's number 7 if you're following along with us at home.) To make the important point that we humans get our energy to grow and move from our food, we planned to use a miniature car that was to be powered, not with batteries or gasoline, but with a steam engine—a steam engine whose heat would be provided by the combusting of sugary breakfast cereal, specifically Frosted Flakes. The chemical energy released by burning the cereal is much the same as the chemical energy released when your body breaks it down. The big difference between a flame and your stomach is the temperature. Your body uses amazing enzymes capable of combining sugars with oxygen at the moderate temperatures of biology. In regular combustion, a chemical reaction releases heat energy much more quickly, which produces the fiery temperatures we all recognize in a candle flame, a rocket's exhaust . . . or a steam engine.

As we were preparing for the episode, the question was raised: What should a Frosted Flake–powered Digestion Car of Science look like? After just a moment of consultation, Bill Sleeth, the set designer, and I agreed that such a vehicle should look, and I'm quoting the both of us here, "Just like the real ones." Generally speaking, of course, there is no such thing as a Frosted Flake–powered car. Nevertheless, we all immediately knew what we were describing: a machine that displayed its functional mechanisms openly and honestly. The viewer would see the technology we used, not some kind of whimsical cartoon version of a car that kept the working parts hidden. It would have been easier to build an electric, remote-controlled toy car carrying a dish of stand-in cereal, but then I would have been forced to tell the viewer something like, "This is a representation." That's not how we did it.

Bill Sleeth and his guys ended up creating a hobby-size steam engine mounted on Erector Set parts. You could run your eyes across our Digestion Car and tell exactly where the combustion happened, where the heat from the burning flakes created steam, where the steam moved a piston, where the moving piston turned the wheels. It was a model of nerd honesty. Yes, we could have made it prettier and shinier, but then the car wouldn't have been nearly as entertaining. It was entertaining because it advertised its function and invited you, the viewer, to understand exactly what was going on. Bill's team pulled it off perfectly. All the while, those same two Disney executives were shaking their heads at us country bumpkins and our silly ideas about trying to do it all as though we were *really doing it*. But we did, and it paid off. My sources assure me that *The Science Guy* show remains the primary income generator for the arm of Disney called Disney Educational Productions, and I maintain that the nerd honesty embodied in the Digestion Car is the essential reason for its success.

Reflecting on the beautifully naked functionality of our little contraption has helped me tremendously in many different parts of my life. I hope that kind of authenticity will help you, too. Certainly it would have helped many of the engineering bosses and television producers I've worked with over the years, not to mention a lot of the politicians and would-be scientists I've encountered. That thought inspired me to write up a short guide to keeping it real.

These may seem like sweeping lessons to take from a balloon, some bricks, and a cereal-eating car, but just imagine if more of us lived by those principles. It would mean not having to wonder whether a salesperson you're working with has an ulterior motive, not worrying that a contractor is cutting corners just to finish your job faster, not having to question whether someone is telling you the truth or skewing toward some personal agenda. It would mean being honest with your audience, customer, engineer, whatever, and, above all, with yourself—not just some of the time but all of the time, even when cutting corners looks oh so tempting. It would mean thinking through each step of the process to the best of your ability, stretching and learning as you go. And then carrying out your plan with nerd stubbornness until you get exactly where you want to go, whether conventionally or cereal-poweredly.

With that in mind, here's my Nerd Code of Conduct:

- Be open and be honest.
- Don't pretend you know what you don't know (often a little too easy to do).
- Show the world as it is, rather than the way you wish it would be.
- Respect facts; don't deny them just because you don't like them.
- Move forward only after you trust your design.

CHAPTER 15

Resonating to the Nerd Beat

For me, television has been a wonderfully effective way to break down barriers and to help spread the nerd worldview. That is why I abandoned my life as a full-time engineer. When I walked off the stage after performing as Bill Nye the Science Guy for the first time, on January 22, 1987, I was pretty sure I had made the right decision. I felt as though I had "killed it," as we say in comedy. Aside from my tongue still being a little cold from chewing marshmallows frozen in liquid nitrogen, it seemed natural, even. Much as I loved getting laughs, though, I wanted to fold in more real information. My short *Almost Live!* bits were really short and more about laughs than ideas. I could sense the potential for a fun, educational television show hosted by none other than . . . me. So I decided to discuss my career with someone who knew a whole lot more about hosting science TV than I did: Carl Sagan.

Because of good dumb luck, I had taken an astronomy class with Professor Sagan when I was an undergraduate at Cornell. This was 3 years before his famous *Cosmos* series aired, but he already had the passion that made him a compelling lecturer. By 1987, he also had the experience and knowledge of how to talk about science on TV. I figured, "Man, this guy is the expert of experts. I've got to talk to him about my idea for a career." Why not? My 10th college reunion was coming up, and I was going to be in Ithaca, New York, anyway. I contacted his office, and I eventually managed to convince his secretary to schedule 10 minutes with Professor Sagan. I told him about some television segments I'd been working on called *Bill's Basement*, and described my vision of the much greater things I wanted to do as the Science Guy. He listened thoughtfully. He said that he liked my concepts on the whole but advised me to avoid engineering demonstrations and to instead focus on pure science. His explanation was as succinct as it was memorable: "Kids resonate to pure science."

"Resonate." That was the verb he used. "Resonate" is a wonderful word that appears in a great many different disciplines. Professor Sagan had talked about resonance from time to time in class when he was describing the coupling of a planet's orbit with its spin. The Moon spins exactly once during each orbit around our Earth. The motions of the two bodies are locked together in synchronous resonance. When you ride a swing or give a kid a push on a swing, you do so in synchrony with the swing's motion. You add energy at its resonant frequency. Resonance is what happens when objects vibrate at their natural rhythms, so that they respond strongly to small impulses. It is also how we make music. When you pluck a guitar string, blow into a flute, or crash a stick on a cymbal, you are putting energy into the musical system—the string, air, or metal—at just the right rate to activate the material's natural tendency to vibrate. You can produce a lot of sound with a small amount of breath or motion. You pluck the beauty of resonance right into thin air. But here he was, Professor Sagan, talking to grown-up young me about resonance in yet another, quite different context.

I heard a powerful metaphor in there. Educators embrace the idea that a single lesson delivered in just the right way will resonate with kids, producing a lifelong change in them and the way they think. It is the dream of every teacher and every parent. It is something that television might be able to do, too, I thought. It seemed reasonable, since I had experienced it myself. In high school my buddy Ken Severin and I had watched *Frames of Reference*, a film about inertia and motion, presented with spot-on wit by Drs. Hume and Ivey, over and over again in the physics lab after school. Incidentally, I've long since let go of the dream of doing anything as good as *Frames of Reference*. Nevertheless, I carry on. I aim for resonance in all my talks, and I aim for it in this book, too.

By itself, a film, show, book, or rally is just a small, good thing. But if a TV show is done well and really connects with its audience, I realized (and hoped), its influence can grow to be enormous. In technical terms, I might say that the amplitude of the resonance can become far greater than the amplitude of the forcing function. In regular terms: Small science lessons—delivered with enough humor, energy, and empathy—can do more than change minds. They can potentially change the world. So after a pep talk from Carl Sagan, that is pretty much what I set out to do.

I thought a lot about how to capture the essence of nerd thinking and the scientific method. I poured all those ideas into the television show *Bill Nye the Science Guy* on the Disney Channel. The show won 18 Emmys and is still shown in schools, so I guess things worked out okay. Professor Sagan's insights opened my mind, my show opened a lot of other minds, and today the process is still going and going. I like to think kids still resonate to the science the crew and I showed them. I hope that my fans, in turn, are doing a lot of resonating of their own.

❯ ❯ ❯

Carl Sagan's whole career is a fascinating case study in nerd resonance. He is probably most famous for his 1980 television series *Cosmos*, which vibrated off his popular science writing and television appearances of the 1970s, which built on his planetary research of the 1960s, which began with his scientific passions as a student at the University of Chicago in the 1950s. The 1960s were the height of the Cold War, when it seemed as though the world's nuclear powers were about to start a nuclear war in response to a random provocation, or even by accident. Yet not long after Nikita Khrushchev was warning the United States, "We will bury you," Sagan was working hard to engage Russian and Soviet bloc colleagues in scientific collaboration. While much of the public's focus was on missiles and warheads, he was directing our attention to the worlds of the solar system. The United States landed the twin *Viking* spacecraft on Mars. The Soviets landed the *Venera* spacecraft on Venus. Planetary scientists on those missions shared visions that transcended national boundaries. Sagan especially was vibrating with everything all at once, connecting seemingly disparate ideas across the disciplines of science.

One of the hottest topics in planetary science in the '50s and '60s was the study of craters. Researchers had only recently converged on the idea that most, if not all, craters on the Moon were caused by asteroid impacts, not by volcanoes. When you look at the Moon, you see that it is pockmarked everywhere, so you infer that an enormous number of impacts happened in the primordial solar system. But if that's true, where were the impacts on Earth? When you look at our own planet, you don't see many craters at all. But then scientists realized that we weren't seeing the full picture. Earth's surface has a lot more going on than the lunar surface does. Earth has an atmosphere. We've got rain and snow and wind, and with those come erosion and weathering. Most importantly, the surface is continually being reshaped on a global scale. Geophysicists at the time were just starting to figure out that Earth's crust is comprised of enormous slabs that came to be called tectonic plates. Their motions are driven by slow, powerful forces deep inside the planet.

Plates move exceedingly slowly, at about the rate your fingernails grow, but given enough time, the movement of Earth's crust does an excellent job of erasing craters. Tectonic plates grind against each other. They slide under and over each other. They trigger volcanoes and drive up mountain ranges. Even if these things don't make craters vanish entirely, millions of years of weathering render them nearly impossible to recognize. The Moon and Mars don't have plate tectonics, and Mars has only a wisp of an atmosphere, so once an impactor impacts, the scar lasts a long, long time. Here on Earth, craters fade away.

Carl Sagan and his contemporaries, including early crater experts Gene and Carolyn Shoemaker, got to thinking about how many impacts there must have been on Earth over its 4.5-billion-year age. The Moon offered some clues. In 1965, NASA's *Mariner 4* space probe flew by Mars and showed a shocking number of craters there, too. Earth is a bigger target and has more gravity, so it probably has been hit more often than Mars and far more than the Moon. Sagan, the Shoemakers, and others wondered about the effect of a major asteroid impact here. It would kick up colossal clouds of dust. The enormous amount of heat unleashed by the energy of the primary impact and by all the secondary material it ejected would trigger a worldwide firestorm. All that smoke and dust would block out sunshine and cool the planet for many, many years afterward. It would be a climate catastrophe.

In 1977, when I took Professor Sagan's astronomy course at Cornell, he talked to us students about a new line of research he was pursuing. Asteroid impacts are not the only thing that could catastrophically alter the Earth's surface, he noted. The detonation of nuclear weapons would also cause huge environmental disruption. In collaboration with atmospheric scientist James Pollack, Sagan had worked up a computer model predicting what would happen to Earth's climate in the event of a full-scale nuclear war. The result of Sagan and Pollack's simulation looked eerily familiar: fires, enormous clouds of debris and dust, and then a long cold spell. The consequences were not unlike those of an asteroid impact, but a nuclear war between the United States and the Soviet Union seemed a lot more likely.

Sagan and Pollack called the phenomenon they discovered "nuclear winter." Sagan described it in detail for us in class. He wanted to convey to us the power and significance of computer models and how different scientific ideas can connect. I believe he also wanted to impress on us the importance of a scientist's responsibility. If we wanted to avert nuclear war and nuclear winter, we had to do something about it. Both parts of his message reverberated and made a deep, lasting impression on me.

There was another, quite different, er, fallout (sorry) from all that study of nuclear winter and asteroid impacts. In the late 1970s, the father-and-son team of Luis and Walter Alvarez—physicist and geologist, respectively—were looking for chemical clues about how quickly the ancient dinosaurs died off at the end of the Cretaceous period. Along the way, they discovered an intriguing geological layer enriched in the element iridium. It shows up in rock layers of a particular age, all around the Earth. That was odd because you wouldn't expect to find iridium anywhere near the surface. It is a very dense metal, twice as dense as lead. When Earth was young and molten, geophysicists presumed, almost all the iridium must have sunk toward the planet's core, far below the crust. But studies of meteorites had revealed they often do contain quite a bit of iridium. Meteorites are generally too small to have stayed molten long, and they generally didn't have enough gravity to sort material by density the way Earth did. The Alvarezes reasoned that their layer of iridium couldn't have come from Earth's insides, so it must have been deposited here from the outside . . . by an asteroid.

Here comes the really exciting part: The iridium layer shows up in rocks that are 65 million years old, from exactly the time when the ancient dinosaurs went extinct. It is strong circumstantial evidence that an asteroid impact triggered a mass extinction and wiped those creatures out. The discovery answered a long-standing question about what happened to the ancient dinosaurs. I found that answer very satisfying, particularly in the face of other leading theories of the time. When I was in 2nd grade, my teacher, Mrs. McGonagle, read to us from a big book claiming that the ancient dinosaurs died out because some mammals took all their food. Even Mrs. M recognized that explanation was pretty lame. A *Tyrannosaurus* getting her or his lunch money stolen by a sort of proto-rabbit? Seemed more likely to me that *T. rex* would squash the bunny the way an elephant might smite an ant. Starting from the evidence of that iridium layer, we now have a vastly better explanation for what happened lo those many years in the past.

The Alvarezes realized that an impact big enough to blanket the whole Earth with that much iridium would have left a gigantically huge crater—so big that some trace of it should still be visible today, despite the dynamic Earth's great obscuring powers. If they could find that crater, they knew, it would greatly strengthen their theory of what happened to the dinosaurs. Walter Alvarez had started out as a geologist in the oil industry. Petroleum geologists routinely use magnetometers (which are like a sensitive compass) to map buried geologic structures. Some chemist colleagues published data that revealed traces of iridium in a huge crater, which is mostly submerged and buried in sediment. All the evidence links the long-ago extinction to that 180-kilometer-wide crater, called Chicxulub, which lies along the coast of the Yucatán Peninsula in eastern Mexico.

So the story went like this: Studies of the Moon and Mars showed that asteroid impacts must have had a major influence on our planet. Building on that idea, Carl Sagan started talking about nuclear winter in a "listen, people, we have gotta get our act together" kind of way. Luis and Walter Alvarez really made his point by showing that an extreme version of nuclear winter, caused by a huge asteroid impact, seemed to have devastated the planet and wiped out the ancient dinosaurs. Plate tectonics had erased most of the evidence, but a layer of iridium proved that it happened anyway. And a big hole in the ground near Mexico was revealed to be the scene of the whole event. That's a pretty compelling story of scientific resonances.

In my world, things are still resonating. Many years ago, in 1980, I joined The Planetary Society, partly due to the class I had taken with Carl Sagan. In 1983, the society sponsored a field trip to Belize, to the rim of the Chicxulub crater, to collect rock samples; recently, a group of researchers asked to look at those samples and reanalyze them. Maybe they contain more clues about the mass extinction 65 million years ago. Today, I'm the CEO of The Planetary Society that Professor Sagan cofounded. The society's sunlight-driven spacecrafts, *LightSail*®*-1* and *-2*, are directly based on an idea that Professor Sagan introduced on *The Tonight Show*. And now I advocate for asteroid detection and deflection missions to keep us from getting wiped out by a rock from space, like those poor ancient dinosaurs. Each idea leads to a deeper understanding of our world, which leads to more knowledge about how we can protect and improve our world. The work of scientists, explorers, and researchers of the past resonates in the work that we undertake today, just as our actions shape what future generations know and do. No pressure.

◈ ◈ ◈

In 1993, I wrote a kid's book called *Bill Nye the Science Guy's Big Blast of Science*. It includes an explanation of the greenhouse effect, in which I compare Earth with the closest planet, Venus. On Earth, the global average temperature is about 15°C (58°F). On Venus, the average temperature is about 460°C (860°F). Venus is closer to the Sun, but that doesn't explain the drastic difference; it gets twice as much sunshine, but it is also covered with clouds that reflect twice as much energy back out to space. What really sets Venus apart from our planet is its atmosphere, which is 90 times as thick as Earth's and made almost entirely of carbon dioxide. All that CO_2 produces a super-greenhouse effect, and as the result is a world where even the coolest day would melt a lead fishing weight into a puddle.

Comparing Earth with Venus is a pedagogical path that Carl Sagan took us down when I was his student 2 decades earlier. Sagan and climate scientist James Hansen realized that the greenhouse effect explains the other planet's extreme temperatures. Later they connected the Venusian studies to the possibility of climate change on Earth. As with asteroid impacts and nuclear winter, Sagan found the connection between two seemingly disparate ideas and brought a distant discovery home.

I've been fighting the climate-change fight for more than 23 years now, along with many others out there, the full-time climate scientists. Hansen, the former director of the NASA Goddard Institute for Space Studies, did an early study conclusively showing that carbon dioxide

produced by human activity is making the world get warmer faster
than at any time in the past few hundred thousand years. Michael
Mann at Pennsylvania State University produced the famous "hockey
stick" graph illustrating the world's temperature over the past several
thousand years. Earth's overall temperature was steady for millennia,
but now—whoosh—it has shot up swiftly in just the last 250 years.
Gavin Schmidt, who succeeded Hansen at the Goddard Institute for
Space Studies, advances and refines our climate models by the day. But
in a world filled with people who willfully promote misinformation to
further their own agendas, somehow it is still a fight to get this reality
taken seriously. Does that mean we should give up? Should we sit back
and let those people destroy our future so that we can say, "I told you
so"? Of course not. What it means is that we need to embrace the Sagan
approach even more vigorously. Take the long view. Be resolute but
remain optimistic. Relate ideas in ways that people understand through
clear storytelling and personal connections. Look for ways to build
small actions into bigger effects. Be a regular force that resonates.

What I want to do is get everybody in the United States, everybody
in the world, on board with the exciting opportunities. We don't want
our greenhouse effect to get away from us. As Sagan warned, we do not
want to be like Venus. We can produce clean electricity in new ways.
Sunlight and wind energy can be harnessed with existing technologies.
There's tremendous energy in the primordial heat of Earth, which we
can access just by drilling down a few hundred or even just tens of
meters. And here's something that's worth remembering: You can't
outsource the erection of a wind turbine tower. You can't install a
power transmission line's support towers anywhere but where the
transmission line is. The jobs to create the renewable economy would
be here on native soil. We've seen a recent eruption of populist politics
around the world, propelled in part by complaints about the loss of

local control of the economy. Well, if you want locally produced energy, you are not going to do better than wind, solar, geothermal, and tidal energy. It's yet another instance in which the best, nerd-certified solution ends up benefiting everyone.

A year after I started *The Science Guy* series, Sagan was diagnosed with blood cancer. He died 2 years later. It was a huge loss for the world. He was a passionate advocate not just for science but for the scientific way of thinking. He was the nerd's nerd, but he also had a disarming common touch, an approachable quality that would draw you in when he was chatting with Johnny Carson. People wanted to hear what he had to say. They naturally resonated with his ideas. I'm sure he would have continued to be a powerful voice motivating people to step up and deal with climate change. We could use his help right now, but this resonant process doesn't depend on just one person. It depends on all of us.

Part of the reason for this book is to enlist your help in making that happen—getting you to be part of the resonance. Help people link extreme weather events with the global warming that makes those kinds of events more likely. Help people understand that renewable energy comes with local control. Connect the inspiring discoveries of space exploration with the things we now understand about the danger of rapid climate change on our own planet. Making these conceptual and personal connections is a very Carl Sagan way of communicating the science. I'm sure it works, because I've seen it work.

» » »

One of the biggest roadblocks to getting the message out is doubt. Here, too, I draw inspiration from Carl Sagan. I often hear people use the words "skepticism" and "denial" interchangeably (especially among those who refuse to acknowledge climate change), but these two words are worlds apart. Skepticism is a discipline. It's a component of critical thinking that helps to keep you from fooling yourself or allowing yourself to be fooled; Sagan wanted us all to carry a "baloney detection kit." Denial, on the other hand, is like a lock on the tool kit that keeps you from thinking about ideas you don't like.

The professional denial business pretty much started with cigarette companies. Tobacco companies hired scientists who were careful to say that they couldn't be 100 percent sure about anything having to do with cancer and cigarettes. They explained that it's their business to develop hypotheses and continually question accepted truths. The denial community exploited that aspect of scientific inquiry in an amazingly deceptive way: If there's a 5 percent chance that your friend's lung cancer was not caused by cigarettes, they'd say, then you can't *prove* that cigarette smoking is dangerous. They twisted nerd honesty to their own ends. The underlying message was: focus only on the doubt and don't think too much. It is a way to prevent people from making otherwise obvious connections.

There must always be uncertainty in science, so there are always these kinds of vulnerabilities. It's not just climate change; there's also the opposition to vaccination, the irrational fears about genetically modified organisms (GMOs), all the way up and out to those on the fringe who think the Moon landings were faked. The way we will get past all these manufactured doubts is by exposing people to some important critical-thinking skills. In this example, consider how difficult it would be for landing-fakers just to create all the paperwork that the space program generated. That alone would have been more trouble than landing on the Moon! Critical thinkers can recognize when they are being deceived or manipulated. I urge you to promote critical thinking and practice it every chance you get. Set it out into the world and help it spread. I was lucky to grow up in a family that treated the scientific method as the normal, everyday way to think through problems. Not everyone is so fortunate.

Perhaps it was easier during the Cold War, when Carl Sagan was researching and I was growing up. Back then, embracing science felt like a matter of life or death, because research led to remarkable new weapons. A sense of crisis forced people to put differences aside. Today, science and technology are more advanced than ever, but many of us often fall into the traps of apathy and doubt. It falls heavily on us nerds to rekindle a sense of shared purpose and common benefit. Commitment to spending on science and technology requires public support. Progress requires collective investment of our intellect and treasure—brains and money. This is the downside of resonance: Destructive impulses from the top down and from the bottom up can shake the whole system, too. In the 1980s, the Reagan administration sidetracked us scientifically. The president symbolically removed the solar panels from the roof of the White House. His people did not take AIDS or acid rain seriously at first. They gutted vital basic research programs.

The current antiscience and antiexpert movement has been decades in the making. Now we need to work hard to turn things around. We nerds all have to use our big-picture point of view to make those crucial personal and conceptual connections—to amplify our influence whenever and wherever we can. We have to write, speak, teach, publicize, and do anything we can to resonate with the wider world. For me, the most profound discovery ever made by us humans is this: You and I, and everything we can touch and see, are made of the same materials and driven by the same energy as everything else in the universe. We are one with each other, with our planet, and with the cosmos. We all resonate to the same beat, and if we put something beautiful into the world, it can spread and grow.

CHAPTER 16

Critical Thinking, Critical Filtering

When I was a kid and I wanted to look up an odd or obscure fact or a piece of information, like Millard Fillmore's political party affiliation, I hit the books—the actual paper books—in a library. Or in high school if I wanted to know the atomic number of rubidium, I looked it up in the *Encyclopedia Britannica*, or if I was feeling hardcore, the *Chemical Rubber Company Handbook of Chemistry and Physics* (the ol' *CRC*). Nowadays I just pull out my laptop or my fancy new phone and go to Google. Then 586,000 results and 0.46 seconds later, I learn that Fillmore was the last president affiliated with neither Democrats nor Republicans, lucky guy. And 146,000 results tell me that rubidium, symbol Rb, has an atomic number of 37, which means that it contains 37 protons. Checking the density, I found both 1.53 grams per cubic centimeter and 1.532 grams per cubic centimeter.

Those are a lot of data to absorb, and they are just a tiny portion of what came up. Meanwhile, I had to decide which details mattered to me and which sources I trusted. What about those two different values for density—is the first number rounded from the second one? If you were so motivated, you could dig through a dozen sources, see who did the original measurements, and probably figure out who then rounded it down. In the old days, you had to look things up in just a few reliable sources to save time. In today's data-soaked world, though, you easily can do quite a bit of extra sleuthing. Information comes at us so quickly now that the challenge is not speed and efficiency but figuring out which of those 146,000 results contain the highest-quality answers. Although you and I largely trust the Google search algorithm to put the good information up high, we don't necessarily expect Google to put the best information up at the very top. And how much do you know about the workings of that search algorithm, anyway?

Information is the currency of the nerd perspective, so the rush of online info is enormously empowering. But yikes, does it have some troubling features. No search algorithm is foolproof, especially since people are constantly trying to game the system, hoping to take advantage of a version of a mob mentality. YouTube videos from hucksters and hoaxers can appear as slick as those from major academic and government institutions. And just look at the mess on Facebook, where friends' comments, strangers' comments, sponsored content, legitimate news, and commercial propaganda mingle freely—and confusingly. With so much sense and nonsense flying around, who can sort out what is true and what is not? We need a sophisticated filter that lets through only high-quality information.

A top-down approach to filtering, with someone or some system deciding what is or can be posted online, is not a good answer. Restricting access to information in any way would almost certainly be a terrible development. I wouldn't trust that approach even if the nerds were in charge, and it's a good bet that they would not be. Governments could simply blot out news they don't like, in the way that the Soviet Union tried to suppress details about the Chernobyl disaster in 1986 or the way the North Korean government still restricts most Internet content. Whole classes or parts of the world might be cut off, able to connect only with what looks like real information but is actually false. It would take the problem of information quality to an even higher level.

The real solution—the only meaningful solution—is to be able to evaluate the data yourself and tackle the problem from the bottom up. In the 21st century, data-filtering literacy is therefore as vital as science literacy. Or you can call it by the more common term "critical thinking." Either way, it is a defense against information overload and a powerful technique to overcome confirmation bias. Carl Sagan referred to a critical thinker's checklist as a "baloney detection kit." For me, there are three key ideas in thinking critically about a claim that you come across. First: Is it specific? Second: Is it based on the simplest interpretation of the phenomenon? And third: Has it been independently verified? You could even boil it all down to a single statement: "Prove it!" Let me show you what I mean.

Start with specificity. A meaningful claim should be precise enough that we can agree on what we're talking about. If it passes, it is something we can test. If it fails, if it's not specific, we can stop right there. A vague claim is going to be either useless or wrong—guaranteed. Here's the classic example: Someone tells you, "We live on a great big ball." Put aside what you've been taught; why should you believe it? I mean, that is an extraordinary claim. A big ball? Really? If you go walking around in your neighborhood, you'll see that things look pretty flat in the big picture. Even if there are large mountains nearby, Earth still looks like a flat world with some big and small bumps. It certainly does not look like a ball. If you look out toward the horizon, you might conclude that the world is flat but very large. Anything beyond the horizon is just too far away to see. The sense of flatness is especially strong if you've ever stood on the deck of a ship surrounded by ocean. The horizon really does look like it is just a faraway edge of a flat disk of Earth. Based on those kinds of direct personal observations, it does not seem likely or even reasonable that the Earth is round.

But the claim that the Earth is round is succinct and precise. We can easily test it. We can observe the Earth's shadow on the Moon during an eclipse. We can watch ships sail over and below the horizon, then turn around and return to port. We can build spacecraft and take pictures from high above the Earth's surface. "We live on a ball or sphere" is a testable statement, and we can prove it in a number of ways. The proofs are so straightforward that Greek natural philosophers deduced the roundness of the Earth more than 2,000 years ago. Despite pop-culture myths, educated people in Western culture have accepted the roundness of the Earth ever since. (It's another piece of modern misinformation. The idea that many or most people once believed the Earth is flat is a specific claim that is easily debunked by looking at writings from the Middle Ages.)

Next, let's consider the principle of simplicity. This is tied to what is often cited as "Occam's razor." It's the idea that a simple explanation for a phenomenon is more likely to be correct than a complicated one. William of Occam was a 14th-century English philosopher who advanced this idea as part of a broader argument against abstraction and complexity in the world. It's a reliable road to reason. Try this: "I got a call from my dead aunt today. I saw her number on my caller ID, but when I answered the phone, there was no one on the line. I think it was her ghost." Well, it *could* be that dead people leave behind invisible ghosts (which nobody has detected) and that ghosts can make phone calls (for no knowable reason) and that when they do, they trigger the caller ID of the number they used when they were alive. Or it could be that the phone company assigned your dead aunt's number to someone else, and when you answered, the stranger with the new number hung up.

Which explanation is more likely? I like this example because it is drawn from real life. Someone tried to sue a skeptic friend of mine who debunked this very claim in this very fashion.

Occam's razor is also excellent for cutting down (sorry . . .) conspiracy theories. If someone tells you that doctors, scientists, pharmaceutical companies, government agencies, and journalists are all working together to cover up the dangers of vaccines, think about the complexity of the coordination needed, and all the cover stories required, to make that work. Think about the motivations of all the people involved. Then consider the alternative interpretation: People naturally look for explanations when things go wrong with themselves or with their children, and vaccines—something injected right into the body at about the same time in life as a diagnosis of autism is reached—are an obvious place to project fears when there is no obvious cause. Which explanation is simpler? Which one just makes more sense?

Finally, there's the matter of testability. Specificity and simplicity help get you there; without them, you don't even know what it is you are trying to test. But not all things simple and specific are true, so claims that make it this far still require verification. I'm pretty sure you don't have the time and resources to test every reasonable-sounding piece of information on the Internet. Fortunately, many other people have already pioneered the "prove it" process for you; all you have to do is follow their trails in an attentive manner. Even Wikipedia, the busy (or lazy) person's encyclopedia, is full of references and lists of related sources (60 of them in the entry for rubidium, for instance). Information that has no source is immediately suspect because you cannot know how or if it was tested. When you do see sources, see if they point back to a primary journal, textbook, or researcher at a major research institution. At that point, you can start to rely on expertise—that is, on the people on the high side of "everybody knows something you don't." As with the examples I mentioned at the beginning of the chapter, the Internet can make this vetting process quick and easy—once you practice it a little.

With this said, we are living in a weird time for critical thinking. Climate change is a prime example. Several decades ago, scientists started seeing indications that the whole world is getting warmer overall. Since then they have gathered enormous amounts of data to verify and quantify the discovery. The claim today is quite specific: Earth's temperature is rising, and industrial emissions are the primary cause. It is testable, and nearly all climate scientists will tell you that the evidence for human-driven global warming has in fact been tested and thoroughly verified. Yet a determined collection of climate-change deniers has managed to sow doubt here at the testability stage. They question the researchers' motivations. They question the quality and quantity of the evidence, implying (incorrectly) that there is not extremely strong agreement within the climate-research community. That is why some scientists and science journalists push back, noting that there is about a 97 percent consensus that humans are driving climate change. Their point is not that a mob must be right. It is more an appeal to Occam's razor. It would take quite an elaborate conspiracy to get that many people to sign on to bad or crooked results. The far simpler explanation is that the researchers are doing exactly what they appear to be doing, gathering the best-available data and subjecting it to the best-possible analysis.

None of the climate counterclaims seem worthwhile to me, but I take the need for critical thinking seriously. This is a great opportunity for you to apply the standard of "prove it" for yourself. I think it is worthwhile to work through how you even know such a basic fact as the roundness of the Earth. So by all means—when it comes to climate change and global warming, I encourage everyone to evaluate the preponderance of evidence and to examine the publications by climate experts. As a critical thinker, you are like a juror in a very important trial, perhaps the most important one ever. The case here is one that will determine the welfare of billions of people.

Have at it, my fellow nerd!

Testing ideas is, so far, the best system humans have ever come up with for building an honest, progressive understanding of the natural world. It is a very effective way of getting around the inherent biases and gaps in the way our brains work. Philosophers will tell you that you absolutely cannot trust your own senses. This goes all the way back to René Descartes and *cogito ergo sum*—I think, therefore I am, or the only thing you can really be sure of is that you are thinking. Everything else is subject to question; seeing is not necessarily believing. That concept may sound huge and vague, but it has very real and specific consequences. It is how magicians make a living. It is why science demands repeated observations, independent verification, and falsifiability: An idea can be valid only if there is some logical way to show that it would be invalid. If the Earth were flat, why can't I see Australia from California? These different kinds of tests are all designed to weed out subjectivity and dishonesty. We also have instruments that can give objective information about the natural world, although that information still requires human collection and analysis.

I'd like to see a society in which everybody understands how the scientific method works and why it is so important. Most people will never use it as professional scientists, but everybody should have access to that crucial part of the nerd tool kit as part of their daily process of information-filtering. Despite one education reform after another, we're still not there. Science teachers are quite adept at teaching students how to evaluate the quality and reliability of data and how to recognize errors or fraud. To a large extent, we have the right mechanisms in place. That's an important start. Putting those mechanisms to work requires getting students into classrooms with competent science teachers and expanding the scope of their curricula. But still, that is generally not enough, because it takes time.

Learning to recognize hoaxes, scams, pseudoscience, and the like takes practice. For most of us, it goes beyond the standard school curriculum. It requires a more advanced everything-all-at-once training. One wonderful teaching example is the Web site dedicated to the rare Pacific Northwest tree octopus. This remarkable creature lives in just one small section of the temperate rainforests of Washington State. The tree octopus pounces on frogs or rodents from tree limbs and then crawls back up to safety. It has a mucus coating that protects its body from drying out. It is a true evolutionary oddity: Its aquatic ancestors were isolated on land when the oceans receded in this area, and one isolated population adapted by developing a unique tree climbing ability. Truly a biological marvel. The site even encourages Web visitors to become a "friend" of the Pacific Northwest tree octopus because the poor creature is endangered—what with all those mean-spirited, forest-denuding loggers, and all. The site issues a call to arms: "Together, we have the power to build a grass-roots campaign to save the Tree Octopus." It adds an additional emotional element to the story, and it is very effective.

What? You've never heard of the Pacific Northwest tree octopus? Good. That's because it doesn't exist. It is a hoax created by someone who calls himself Lyle Zapato, a devious prankster who has created a marvelously detailed and convincing Web site dedicated to this imaginary creature.

In one test, educators in Connecticut asked a group of 25 seventh graders to look at the tree octopus Web site. Every single one of them accepted it as real. At the time of the test, the students all went to the same fake Web site, found the same fake information, and compared fake notes with one another. Then they all concluded that the fake "facts" were real. They didn't pursue multiple reference sources. They did not filter the low-quality information, because they had no training in how to do it. Today if you Google "tree octopus," this site is (naturally) the first that comes up. Now at least there is a debunking Snopes page that shows up as well, but even so, it is easy to get sucked into the seductive falsehood of the Zapato site. The story makes for a wonderful science-class lesson—thank you for that, Lyle Zapato (*if in fact that is your real name*!). But for me there's a much bigger message here.

High-quality science education will help vaccinate students against the gullibility that makes the Pacific Northwest tree octopus look credible. And of course, I hope every kid encounters a few passionate, brilliant science teachers on the way to graduation. Still, we all have to recognize that teachers alone cannot deal with the vast scope of the information-filtering problem. There is a society-wide challenge here, and it is on every one of us nerds to pitch in and help.

This inability to think critically is a problem for all of us—kid, tween, teen, young professional, middle-ager, and senior alike. We need to act as ambassadors. We need to impart critical reasoning to our peers, expect it of our parents, and encourage it in our friends and coworkers. Challenge people when they make or repeat an outrageous claim that cannot be logically shown to be false. Too many disagreements devolve into unproductive shouts of "you're wrong" and "no, *you're* wrong." I'd like to see us all offering the nerdier and more mean-ingful responses: Where did you see that? How do you know it's true? What if it's not? These are commonsense versions of the testability standard I described earlier, and they are very powerful.

Wait, I can break it down further. We come across so much informa-tion that we can't possibly put it all through a tight filter. Here are some flags to tell you when you should be instantly suspicious of a claim, even before you start looking for supporting data and references.

- Is it part of an ad or "sponsored content"?
- Does it clearly benefit a specific person or company?
- Does it have no obvious source at all?
- Does it contradict things you've heard before? That doesn't make it wrong, just suspicious.
- Is it something you really *want* to be true? If so, you need to be extra careful.

These are simple tests we can all do every day. They don't require reforming the school system yet again. What they do require is a broad and spreading nerd culture that values honest information—and that scoffs gently at those who do not.

One of the most important skills in information-filtering is cultivating habits that keep the most noxious material from hitting your filter in the first place. Some sources are so untrustworthy that they can't be saved. Especially for those of you who read a lot of news online, I have a simple piece of advice: "Don't bother with the comments section." The places where anyone can spout off about an article or a blog post have become notorious information cesspools, where emotions run high and data quality is low or nonexistent. A journalist friend of mine contacted me recently about my criticism of climate-change deniers. She said, "Did you see the comments? You have to respond right away." I explained, calmly, that no, I really did not. It is consistent with the modern expression "Haters gonna hate." As a rule, I do not respond to the anonymous combatants who are passionately hovering over their keyboards 24 hours a day, ready to pounce on an odd point that really, really bothers them; sometimes they pounce even when there's nothing in the original article that's related to the vitriol they decide to post in the comment section below.

It's human nature to fixate on criticism, but that impulse creates a distorted view of the information landscape. Angry comments create (often intentional) digital noise that drowns out the scientific papers, steady reporting, and serious commentary that inspired the reactions in the first place. The overblown impact of angry Tweets or online comments is a modern amplification of what is often called "selection bias" or the "selection effect." Your brain focuses on the few dangerous-sounding comments, written letters, or remarks rather than on the far more common, neutral responses. That problem has probably been bugging us ever since caveman Og was grunting about caveman Thag back in cave-people days. News articles have always gotten angry responses, whether justified or not. In the print era, the reporters and editors were often the only ones who saw the truly crazy or inflammatory letters. With the rise of the comment section and the magnifying power of social media, a few thousand motivated people—along with a few well-targeted bots—can hijack the public perception of a hot-button political or scientific issue. It's alarmingly effective, and we need to learn how to turn down the volume on the dramatic but probably insignificant squeaky wheels.

Angry comments and hoax Web sites work together to create a

dangerous amount of confusion regarding what's true and how strong each argument is. Some are easy to spot, but others are disguised as academic-style blogs or information resources from reputable institutions. In these sources, there is a different kind of filtering going on— filtering to advance a personal agenda, not to find high-quality information. Climate deniers cherry-pick odd facts and create graphs based on confused or conflated data sources. I cannot help but think of Senator Ted Cruz, a remarkable cherry-picker of temperature data. Writers for the online-only *Patriot Post* sometimes brazenly mix data from one graph with another, essentially making up their own climate data to prove their (fictional) point. I lump these in with the intellectually bankrupt information sources you can safely ignore.

In essence, comments sections are breaking free and getting a life of their own. Instead of venting on Facebook—or, just as likely, in addition to—some people generate plausible-looking stories full of misinformation or disinformation. And I'm sad to say, they're not all as much fun as a tree octopus. YouTube abounds with videos showing evidence of NASA's UFO cover-ups or proving conclusively that the Earth is flat. Climate change has inspired all kinds of sites that advance the denial agenda. We can read them all from our armchairs or phones in a coffee shop.

So let's cultivate critical thinking, not just in our science classrooms, but every day as parents, friends, and citizens. Develop an instinctive mental checklist that saves you from the feeling of overload. Let's find joy in filtering information. Enjoy trivia. Listen and read with a sense of irony and humor. Along with a nerd-sharp focus, you can help us shave off the noise with Occam's razor. We can all learn to evaluate claims and cut to the truth. If a wry, nerdy attitude also adds a laugh to your day, so much the better.

» » »

All this filtering and irony-awareness takes time, and it's creating a need for a big change in the way we interact in the world. So many things happen more quickly today than they used to. And not just finding the atomic number of rubidium; almost every information-related task is far easier than it used to be, and most of our mechanical tasks are more automated than they used to be, as well. Our leisure time is growing as a result—and yet our actual leisure doesn't feel like it's growing at all. What's filling the extra time? Information processing.

Most of us get pretty caught up in processing the parade of emails, text messages, Facebook comments, Instagram posts, Tweets, and so on that come our way every day. A few years ago, I'd often get a phone call at 9:45 a.m. from my beloved predecessor at The Planetary Society, Lou Friedman, asking if I got the email he'd sent earlier in the morning. Nowadays, from other protagonists from other stories in my life, I'll get a text about their email, followed by a phone call to my assistant regarding the text . . . about the email.

I hope you smiled or laughed just now because you know what I'm talking about. "I'll resend my email so it's at the top of your in-box." Argh! Data can fly around nearly instantaneously, but people cannot make instant decisions. Our brains have to process more information than ever before. I sure hope our modern elected officials and their staffs are getting better at managing the overflow. I worry that they are still stuck in the mindset of assigning equal importance to every message—or, worse, paying too much attention to the wrong information. Without an effective filter in a leader's brain, any kind of management is almost impossible.

But once you put in the necessary work—the right nerd-filtering and irony—there is great opportunity for all of us, from the average citizens to the top leaders. We are so saturated by the Internet that it is easy to forget just how remarkable it is. Everybody in the world has, or could have, access to everything—essentially all of human wisdom—as long as he or she has a good Internet connection. With a phone and Wi-Fi, you can browse through the Library of Congress; read the original writings of Galileo, Newton, and Einstein; and explore open-access versions of new papers from some of the most influential science journals. Thirty years ago, nobody in the world would have had the time or money to look at all the things you could look at in the next 30 seconds.

The Internet also goes both ways: You can contribute to it as well as take from it. Until now I've been focusing mostly on the negatives, the junk and anger that people contribute. Fortunately, there are also great riches.

Along with information overload, the Internet enables information *pooling*. People can combine their expertise, and it turns out that when you pose a question to a large group something quite surprising happens. The result is you almost always get a better answer, not just faster than what you could get before but also more accurate and more comprehensive. In essence, you are getting the best of all of the bits of common sense and fragmentary knowledge. I saw a demonstration at a TED event in Long Beach, California, that really made an impression on me. The presenter brought a steer, a great big once-male farm animal, onstage and asked everybody in the audience to estimate the animal's weight. There were about 2,000 people in the auditorium, and they got within 2 pounds (1 kilogram) of the correct answer.

This kind of collective knowledge, often called the "wisdom of crowds," got a lot of hype a few years back, but it actually has proven its value. The demonstration with the steer worked because the measurements that were being averaged were being made by people who each had a lifetime of experience in estimating an animal's weight. By this I mean, we all know about how much we and other people weigh, along with how big and tall they are. Steers and people are made of the same materials (yes, we're made of meat), so our estimates had a pretty good shot at coming out about right. When we averaged the estimates of 2,000 experienced pairs of eyes and brains, the answer came out *really* right.

Put enough people together and their collective information filters become kind of amazing. Wikipedia operates along these lines, with an open system of information flow and a top level of directed filtering. So do a large and growing number of citizen science projects that allow any interested person anywhere in the world to participate in data-intensive research projects. For instance, you can sort through radio signals to see if there is anything unusual that could be a message from an alien civilization (SETI@home); or you can examine images of galaxies and sort them by type (Galaxy Zoo). Just like that TED demonstration, these projects combine commonsense wisdom from many people to produce meaningful answers. Only in this case, the resulting insights are genuinely new. The Galaxy Zoo project led to the discovery of a previously unknown class of cosmic objects, called Voorwerpjes (now there's a great word) associated with active galaxies.

In less formal ways, we are all participating in a wisdom-of-crowds experiment. With more access to knowledge comes more connections, more creativity, and more organization that helps filter and sort information. If you cultivate the impulse to keep learning new things, all of these skills reinforce each other and make our connected world that much more valuable. I hope this will make you feel a little better about all the information-processing you are doing every day. The Internet frees you from a lot of other drudgery so that you are free to expand your nerdy perspective. What it can't do is teach you to make the connections that lead to creative inspiration and world-changing discoveries.

That is why critical thinking and smart filtering are so important. If you can sort through the onslaught of information effectively—if you understand how to exploit all this connectivity and connectedness without getting caught in the web of hoaxes and arguments—you have access to a deeper level of understanding than any of the billions of people who lived before you. That's it, the express lane to everything all at once, and it is potentially available to everyone on Earth. So question what you see there. Use nerd honesty as your touchstone. Weed out tainted and unclear claims. Look suspiciously at conspiracy theories and other too-complicated explanations. Seek out testable, repeatedly verified ideas. Hold on to "prove it." Encourage the people around you to be as critical as you are. Don't tell them what to think; show them *how* to think. You can do this. You can be a critical-thinking citizen of the future.

Critical thinking is a global issue every bit as much as information overload is—or it should be. I often make the case that Internet access should be considered a human right, much like access to electricity and clean water. Nerd minds exist around the world, but the only way to unleash them all is to tie them in to the same information hive and to educate them in the ways of smart filtering. If we succeed—if we can make the Internet into a global commons with effective access for everyone—the Earth will be in a different and better place than it ever has been before. Who knows what ideas will emerge, what kinds of change will be possible? Get connected, and we will see.

CHAPTER 17

A Vaccine Against Deception

Have you ever walked on fire, striding over a bed of coals as if you were invulnerable? I have, and I am not. Have you ever seen four young girls running back and forth through a hallway to escape a ghost, which, apparently, had the ability to turn a light on and off? I have, and it did not. Have you ever been convinced by news commentators that since carbon dioxide is only 0.04 percent of the atmosphere, it couldn't have any connection to changes in Earth's climate? I have not, despite their best efforts, and it does.

Take these as cautionary tales. Even after we have mastered the nerdy skills of thinking critically and filtering out bunk, we are not done. As I hinted in the previous chapter, we are facing twin information challenges these days: both one of *per*ception and one of active *de*ception. There are plenty of people ready to take advantage of anyone with flawed filtering. They are specifically trained to create claims that sound plausible enough to get past lax defenses, and to defeat the unaware with a façade of authenticity. Fortunately, nerds have had a lot of time to prepare for the deceivers. The glut of information may be new, but the tendency of some people to lie and deceive for personal gain certainly is not. Each new information technology provides new pathways to deception. Gutenberg's printing press helped spread anti-Semitic libels; sensationalized 19th-century newspaper stories helped spark the Spanish-American War. But some basic ploys to play on the gullible, like palm reading and faith healing, clearly go back at least thousands of years.

When I wander around my hometowns of Los Angeles and New York, I still see signs for palm readers and psychics on every block, faith healers in "clinics," and aisles full of faux diet supplements in even the fanciest grocery stores, where the (supposedly) better educated among us shop. If we had the kind of critical-thinking skills that could stand up to manipulation, we wouldn't have these things. Neither would we have parents mobilizing against vaccination, nor would we have climate-change deniers holding sway over our government. We all—yes all, even those of us who imagine ourselves above it all—need to sharpen the defensive information filters I've described into something that can stand up to an offensive wave of deceit. The place to begin is by knowing your enemy and understanding how he or she operates.

For fun, I'll start with fire walking. This is a classic example of how charlatans and scammers take advantage of people by exploiting our human tendency to believe rather than to question. I've fire walked a few times for television shows to prove that there is nothing supernatural about it. Here's how a typical fire-walking demonstration goes: The sponsors begin by explaining that your spiritual preparedness is the key. (This preamble is designed to draw in the audience, and also to kill time while the coals ash over.) You will walk barefoot across a wood fire, but your inner something or other will enable you to set your skin directly upon burning logs without injury. It's quite a claim, and it sure looks impressive. After all, we are talking about real fire. As a Boy Scout cooking over an open wood fire, I have suffered a couple of unpleasant second-degree burns. My skin formed blisters in an instant. But here's the significant point: I've had only one very minor burn while walking on fire. That's not because I practiced any of the mumbo jumbo psycho-babble in which I prepared my psyche or called upon a greater power or anything like that. It doesn't sound or seem as though it would be possible, but it is because of a great big little thing called physics.

I can strip away the mystery. Generally, the sponsors of the fire walk build the fire on grassy ground. Twice, in two different television-studio parking lots, I watched as the sponsors brought in fresh sod just for the demonstration. They arranged it in long strips to form a rect-angle, like a picture frame with the fire in the middle. The producers claimed that the sod was there to keep the fire from spreading, which is rather unlikely since the fires were built on asphalt. The real purpose of the sod is to hold water: Whether out on a lawn or in a parking lot, the producers always make sure that the surrounding area is soaked. After the fire has been burning long enough to produce glowing embers, they spread the embers with metal rakes to create an 8-centimeter (3-inch) thick walkway of orange glowing coals. Each fire walker starts on the wet sod or grass, takes a deep breath, and walks over the hot coals. It takes about five long steps to reach the patch of wet ground on the other side of the rectangular fire.

If you don't know what's going on, you are generally amazed. You might be *so* amazed (how amazed would you be?) that you sign up for the sponsor's $4,495 life-coaching course. But let me save you the cash by revealing the not especially well-kept secrets of how you can do the exact same thing using nothing but science.

First, you walk quickly. When you watch videos of people walking on fire, you can see that they're not messing around. They hurry, which means there isn't much time for the heat to penetrate their feet. Second, there is an effective cooling layer of water on the bottom of your (or their) feet that's been picked up from the wet grass; that's the real reason the producers soak it beforehand. The water becomes steam when your feet alight upon the hot coals. It takes a lot of heat energy to drive water molecules from liquid to gas, what's known as a "phase change." While the water is vaporizing, it will not go above 100°C (212°F); this is the same reason a boiling pot of water cannot get any hotter than that.

Third, burning wood is not a good conductor of heat. When you stir a pot of oatmeal or sauce, you probably use a wooden spoon for just this reason. Very little heat flows from sauce to spoon to you. It turns out that heat does not travel especially well from fiery wooden coals to your feet, either. Finally, there is perhaps the most surprising feature of fire walking: The bones and muscles in your feet are basically shoe-size pieces of meat that can soak up quite a bit of heat energy without your skin being much affected. As a result, you can walk quickly across a low fire and your feet hardly even feel warm.

There is one more not altogether insignificant effect. When most of us get ready to step onto a fire, we are a little, uh, anxious. For an ancient evolutionary reason, the blood vessels in our extremities tighten when we are afraid. The medical expression is "vasoconstriction." You literally get cold feet. It takes the fire a moment to warm your feet back up to your standard body temperature. If you were to get a cut or an abrasion during a fight or an encounter with an angry bear, you would not bleed as much because of those tightened blood vessels. Cold feet are good for fire walking.

Now, if the conditions aren't right, credulous fire walkers can get a lesson in critical thinking the hard way. Some people walk too slowly and run out of muscle to soak up the heat. Some walk across a fire bed that is too long and get burned near the end, or accidentally kick coals onto the tops of their feet, where the skin is thinner and more sensitive. People sue when they realize that fire walking has nothing to do with the spirit world or extraordinary mental preparedness. But in my opinion, those clients were complicit because they were not thinking well. They didn't stop to think that the situation is susceptible to analysis and experimentation. They didn't consider that the mystical power of the fire walk had more to do with the physical than with the metaphysical, nor did they stop to appreciate just how cool the physics of a fire walk well done could be. Not to put too fine a point on it, but a true nerd would not get burned.

In case you are wondering, the one time I got burned doing a fire walk happened because the person who built the fire for a TV show got started too late, so the coals had not burned for long enough. I was doing a shoot for *Bill Nye Saves the World*, and we were working on an aggressive schedule. I had to take my walk before the crew went home, which was long before the fire had fully settled. There was a lot more coal and a lot less ash than there should have been, but we had to finish the shoot, so I walked anyway. My feet sank in and absorbed too much heat. I'm sure some unscrupulous life coach would be happy to say that it was a failure of mind over matter. In reality, the matter was doing exactly what it always does. Fire walking is a trick of science, not of mind.

》 》 》

"But Bill," you say, "I was already a critical thinker, and after reading that last chapter, I'm sure I've got this figured out. I don't fall for any hoaxes or superstitions." At least, I like to imagine that a lot of you are saying that. I hear things like that all the time. But I have to ask: Are you sure? Have you never had an irrational feeling about a certain place or a certain date associated with bad luck? Have you ever joked about how it will surely rain if you plan a picnic? We all have problems with critical thinking, is all I'm saying.

Recently, I was visiting longtime friends right after they had moved into a new house in Burbank, California. Their kids were half-joking that there was a ghost in the hallway. I say *half*-joking because the younger ones looked genuinely concerned. The lights turned on whenever they ran through one area of this new house, and they couldn't think of any regular reasonable explanation. So naturally, they concluded that a ghost was responsible, and that the ghost was probably why the previous owners had to leave. They were just kids, true, but they were bright kids, and they sure did jump to the ghost conclusion quickly. The reason, I think, is that there are so many ghost stories in our culture. Even if we know they are fiction, they creep into our consciousness, making the idea of ghosts seem somehow plausible. And their parents, my friends, didn't know what to do beyond insisting that ghosts are not real.

I didn't launch into a diatribe about the absurdity of ghosts. Instead, I wanted to drive home a much bigger point. I wanted them to feel empowered to seek out explanations for anything they didn't understand in their lives, including and not limited to seemingly supernatural occurrences. They were young kids, and I wanted them not to be scared of anything like this. So, I worked with them to develop a hypothesis. I suggested that perhaps there was something wrong with the house's wiring. I encouraged them to consider that a connection might be loose somewhere. Perhaps an electrician had failed to tighten a "wire nut," an internally threaded metal tube held in a plastic cone. I explained that to connect wires in modern houses, you strip the insulation for a few millimeters off their ends, twist the wires together, fit a wire nut over the twisted bare ends, and tighten the nut the same way you tighten a bolt.

Then we tested to see if we could reproduce the phenomenon. I began stepping on the floor in about the same place where the kids had been running. The lights flickered like crazy. I opened a door to the outside on the off chance that an electrician might have connected the lights through another access point in the exterior wall. After a moment, I decided that such exotic hypotheses were unlikely (Occam's razor!). I discovered a second light switch in the same room. I saw that a pile of moving boxes had been stacked higher than the switch's position on the wall. As the kids ran through the hall, the floor flexed; the boxes rocked and tapped against the second switch; the switch turned the lights on and off. We had a testable, reproducible explanation. It took the kids only a moment to see that the mystery was solved, no ghosts required. They got on board with critical thinking—but only after I practically held their hands and walked them to the light switch. I reassured myself, and the parents, that the longest journey starts with but a single step.

In science, we want to develop hypotheses and experiments that show something to be false, rather than a hypothesis or test that only shows something to be true or that a specific expectation has been met. A non-falsifiable hypothesis isn't very useful. For example: If I claimed that this book just instantaneously disappeared and was replaced with a lap-sized, fire-breathing dragon . . . and that just now, the dragon disappeared and the book reappeared—and it all happened so quickly that you could neither see nor feel it—well, that would not be a useful hypothesis. The part about "too quick to see" obviates and tosses out the whole silly thing, as there is no way to prove it false. You might respond, "Well, what if I have a high-speed camera?" I would say, "Even your camera isn't fast enough." You might counter with,

"Well, what if I have a heat detector that can sense the dragon's fire?" I'd say, "The fire dissipates too quickly, even for your sensitive gizmo." I could create responses like that indefinitely. With this in mind, I encourage you to try coming up with your own non-falsifiable hypotheses. It's a fun and enlightening exercise.

We nerds have to make it our mission to encourage those around us to question and test the "evidence" that other people present as fact. We need to teach them the nerd mindset because we need as many allies as we can get. The abundance of supernatural ideas all around creates a self-reinforcing situation: If so many other people believe in irrational things, maybe they are not so irrational after all? In my neighborhood in Los Angles, there are 14 psychics within 10 kilometers (6 miles) of me. A psychic is trained to induce her client to tell her all the facts necessary to piece together a convincing "reading." Once the psychic has the confidence of the client, she can deliver just what he or she wants to hear. The word "confidence" is the root of the term "con man." How do all those people manage to make a living? People pay them. And who pays them? In this neighborhood, we're talking mostly about college-educated, very successful people who probably do not think of themselves as superstitious citizens.

As a counterforce to all the predatory deceivers, several skeptical organizations fight the good fight for science. The Committee for Skeptical Inquiry (CSI, get it?) and the Skeptics Society are two of the leading lights. They seek out specific claims, evaluate them, and publicize their versions of the process that I went through with the kids and the ghost. Both CSI and the Skeptics Society put out magazines that promote the aggressive form of critical thinking needed to immunize us to fakery. Prominent members like Joe Nickell and James Randi have put out fascinating articles and books exposing exactly how psychics do their job. I encourage you to support organizations like these.

Your brain is wired to confirm a previous belief about cause and effect. In other words, we are naturally predisposed *against* critical thinking. Maybe horoscopes truly are harmless, but the analogous biased viewpoint is very dangerous when it crosses over to other claims and concepts. Kids are dying because of parents who have developed an unwarranted fear of vaccines. We need to be part of a system that pushes back against irrational interpretations and guides us toward the ones that reflect the genuine rules of reality. This is a challenge even for nerds, as shown by the various hoaxes, frauds, and published retracted results in scientific publications.

In science, the mechanism for pushing back is a terrific tradition called peer review. To get your research findings published in a mainstream science magazine or journal, you have to arrange for other scientists in your field to read and review it before the publisher will accept it. Peer review is a robust system for weeding out bad procedures, faked data, and other deceptions. Can we import some of that process into everyday life? Sure we can. We can make "prove it" a standard part of the learning method in our classrooms—not just in science but in history, social studies, and literature. We can cultivate a kind of informal peer review in our daily lives. When you come across an improbable or outrageous claim, ask the friend or coworker who you think would know the most about the topic. Let your friends know that you'd welcome them asking you, too. Social media and text messaging make these kinds of interactions fast, easy, low-maintenance. Fire walkers promote a culture of credulity. We can promote a culture of skepticism.

But you can see that being a fully engaged nerd in this way is not going to be easy. Confirmation bias comes naturally to us. It's a shortcut that lets your mind move on to the next agenda item in your life. But our scientific method requires us to come up with hypotheses and to design tests to prove our beliefs wrong. It calls on us to constantly question our assumptions about what is "obviously" true. Only after we work very hard to show that something is not true, but keep confirming it, can we truly trust our answer. We have to walk through a very different kind of fire—an intellectual burn that can be really painful in its own way—to get to the kind of enlightenment that really means something. That is the essence of critical thinking.

It is your job to keep yourself critical despite the constant nudges in the other direction from social conditioning and from your brain's self-deceiving tendencies. Currently there is the continual civic reminder "If you see something, say something." It's designed to keep us vigilant against any errant object or behavior that could signal a terrorist attack. Well, much more likely than a terrorist attack is an attack on you by a con artist. So, if you see something—think something! Think! Is what you're seeing credible? Is the effect a potential charlatan is showing you credible?

Vigilance against ghosts and psychics comes naturally to most of us, but that hardly means you are immune to irrational persuasion. If someone tells you about a remedy "doctors don't want you to know about," that person is almost certainly lying or has been misled. What if someone tells you that a compound in red wine is effective at slowing the aging process? Or that a group at NASA has invented a warp-drive rocket engine that can take you across the galaxy in a few moments and doesn't need any fuel? Or that genetically modified foods cause birth defects? These are all real claims circulating out there, and they all require a more advanced form of critical thinking to evaluate. The baloney-detection standards I described before still apply, but they need to be amped up. The harder someone is promoting a claim, the more vigilant you need to be. Revisit the key points.

■ Start with the claim. It's got to be positive; it can't just raise vague doubt, for example. ("I don't think anyone knows what's going on with the climate.")

■ The claim, or assertion, or statement has to be testable, and the outcome of that test has to be repeatable. Beware of "one study found" claims.

■ Question your own motivations for believing a claim. Watch out for those times when you just want it to be true. (Wouldn't it be great if a daily glass of red wine—even better, two glasses—was the key to a long life?)

Critical thinking is so important for all of us. It's not so that we can smugly feel better than other people, but so that we can all make a better world. One of the most powerful examples is vaccination against disease. Edward Jenner made the first reliable vaccines in the 1790s. Since then, thousands of researchers have protected us from hundreds of thousands of deadly germs. Vaccinations work. They save lives. They work so well that many people have forgotten or missed what exactly vaccinations do. In skeptical circles, we call this the "paradox of protection." If everyone around you is vaccinated but you are not, there is still very little chance that a troublesome germ will find and infect you. You think, "I don't need to get vaccinated." The anti-vaxxers live off this paradox, but a much larger number of people experience it at a lower level. They think, even if there's only a tiny risk associated with vaccination, there's almost no risk in *not* being vaccinated. Or they think, "If I don't bother to get a flu shot this season, it's no big deal." It seems reasonable.

I went to elementary school with a kid who had polio. Let me tell you, people, you do not want polio. Almost everybody in the United States and Western Europe has been vaccinated, so a disease like polio seems like a relic from the ancient past. But it's not that way everywhere. On a recent episode of *Bill Nye Saves the World*, our correspondent Emily Calandrelli went to India and interviewed a young software developer who is wheelchair-bound. He contracted polio before the vaccine was common there. He is thought to be the last guy in the country to miss getting the vaccine. As he and Emily pointed out, there is no anti-vaccination movement in India because you can still see polio victims rolling the streets. The threat of misery from microbes is too real. It is easy to understand risk when it is staring you in the face. It is much, much more difficult when the risk feels remote, as it does in the United States. At that point, critical thinking becomes absolutely . . . well, critical.

From a scientific standpoint, the anti-vaxxers are endangering other people. You have to get vaccinated, and your kids have to get vaccinated. This is provable, substantiated science. The key paper claiming a link between vaccines and autism, on the other hand, has been debunked and retracted. The people who still believe in the link clearly have some reason for wanting to believe, but that reason is not scientific evidence. There is a moral aspect to avoiding deception. I mean, giving in to falsehoods or fantasies just isn't right. Finding honest information sometimes requires being honest with yourself.

When people tell you they can overcome fire with their minds, be skeptical. When people tell you that all the doctors are wrong and vaccines really do cause autism, be every bit as skeptical again. Internalize the guidelines of critical thinking in their basic and more aggressive forms. And remember that if you allow yourself to be fooled, you are probably not the only one who will suffer. On the other hand, if you contribute to the culture of careful skepticism, you are not the only one who will benefit. Let's just start there. The Bill Nye nonsense detection kit. Don't leave home—heck, don't leave anywhere—without it.

CHAPTER 18

Destiny Be Damned— Full Speed Ahead

Sometimes it seems like I was destined to become Bill Nye the Science Guy. There was my father, an experimenter and inventor who called himself Ned Nye, Boy Scientist. There was my mother, an expert at solving puzzles and cracking codes. At my high school, I was exposed to the joys of playing with an oscilloscope and swinging a three-story-high pendulum with my buddy Ken Severin. In my senior year, I submitted a yearbook photo of me hugging that oscilloscope, accompanied by a quote: "With this I could, dare I say it, rule the world." Apparently I had mangled or misremembered a line delivered by the actor Boris Karloff. But no matter—that line stuck in my head. Over the years it morphed into "change the world" and became my guiding principle. It made me restless enough to quit my steady—perhaps too-steady—job at Sundstrand, to walk away from the engineering world entirely, and to gamble with reaching a mass audience on TV.

"Change the world" became my catchphrase on *The Science Guy* show. At dinner events with The Planetary Society, I'll raise a glass of wine and offer: "Let us, dare I say it"—and here my colleagues often chant, "Say it! Say it!"—"Change the world!" I can't help it. And here we are.

If you want to change the world, though, destiny is a dangerous concept. You need a lot of tools to be an agent of change. Nerd honesty is not enough. Add great design principles and a deep sense of responsibility, and it's still not enough. You have to believe in change itself—in the idea that you can choose your own direction in life and influence the future, destiny be damned.

An enormous problem for us in the environmental and engineering communities is an opposing view promoted by many climate-change deniers and their allies in business and politics: They admit that the world is getting warmer, but they insist that there is nothing effective humans can do about it. I'm quite confident that they are merely rationalizing their use of fossil fuel energy and their aversion to new, disruptive (and possibly disadvantageous to them) ideas. Still, a small part of me wonders: What if they are right? What if we as a species are just too entrenched in our environmentally destructive patterns to do anything different? What if we *can't* change the world? As a science guy, I have to at least explore the contrary hypothesis.

I'll start by dispatching the easiest and most extreme version of the argument. There is no such thing as the-future-is-foretold destiny. No way, no how. Nerds do not believe in it. There is no prophetic story that says we have to overheat the planet. Destiny (or fate or providence or whatever term you want to use) implies that future events are recorded away somewhere and can unfold in one way only. We know that isn't true.

First of all, no experiment or observation has ever shown any evidence whatsoever for such a preexisting record of the future—and yes, plenty of people have looked. Second, there is the uncertainty principle in quantum mechanics, which states that a certain amount of unknowability is written into the laws of physics. The uncertainty principle isn't merely about the limits of what humans can know; it is about the limits of what *can* be known based on the structure of physical information. The universe is inherently fuzzy in a way that rules out the concept of a clear, preordained future. And third, there is the matter of the flow of information. In Einstein's general relativity, information from the future cannot flow to the past; otherwise, effects could precede causes and the whole system of reality would fall apart. Destiny requires acquiring information about the future, and everything we know says that is impossible.

However, there are softer forms of destiny, ones that are harder to sweep away scientifically. A great many people do not feel like they are in control of the future—not because of physical law but because they do not believe that meaningful choices are available to them, or at least are not practical or feasible. There are all kinds of reasons people reject the possibility of change. Some of them think "the system is rigged" by powerful people, companies, and government institutions. Some feel trapped by social, economic, or personal circumstances that leave them with no obvious options. The irony is that in American society we also routinely celebrate the scrappy individual who has pulled him- or herself up by the bootstraps, by working hard and making good choices.

Biologically speaking, there is no question that our human behaviors are constrained. Some neuroscientists go so far as to argue that there is essentially no such thing as free will; it's just an illusion that covers up the brain's unconscious decision-making process. On the other hand, every mentally healthy adult has the capacity to understand the consequences of his or her actions. It is part of the legal standard for being fit for a criminal trial. Whether we make use of that capacity is another matter entirely.

For the last 20 years or so, a good friend of mine has come very close to running out of gas in her car, over and over. In psychology this is called "recapitulation," the tendency to repeat behaviors regardless of the consequences. She'll freely admit that she could change; she just doesn't—most of the time. But here's the thing that gives me hope. Every now and then, she recognizes the recapitulating behavior and fills up the tank early. If the consequences of driving around with an empty tank were worse, and if my friend made a genuine effort to break her pattern of behavior, I'm sure she could do it. Look at how my parents gave up smoking when the impetus became strong enough. That's the kind of free will I want us to use. I want humankind to do things in new ways because we appreciate that the consequences of not doing so are too dire. The defining attribute of being human is that we, more than any other species on the planet, exert control over our environment and over ourselves. That is our terrifying and wonderful power.

This is where nerd honesty comes back into the picture, because it is key to unlocking control. It allows us to recognize our biases and quirky perceptions. It helps us understand who we are and how we got that way. With that in mind, I started thinking back to some memorable moments in my path to, er . . . uh, absolutely not destiny. Back in the late 1980s, I was working as a freelance engineer and writing jokes, or at least trying to write jokes. Now and then I'd get a week of work as a stand-up comic. I had enough money that I decided it was time to sell my 17-year-old Volkswagen Beetle and buy a new used car—I mean a used car that would be new to me. I bought a Nissan Stanza that was made during the brand's transition year; it had a Nissan nameplate on the back and a Datsun nameplate on the front grille. Well, it turned out that on the other side of the country, my sister bought the same make and model car, from the same year. Both were four-door hatchbacks, white outside with a red interior. If that's pure coincidence, it's an awfully specific one. Then again, I don't buy the idea that fate intervened in my used-car choice. So I thought back a little more.

When my sister, brother, and I were young, my parents regularly took us on family vacations. To have enough space for everyone, Ned and Jacquie Nye purchased a white 1963 Chevrolet Bel Air station wagon. My father had wanted a white car with a blue interior. The car that had arrived at this dealership was white with a red interior. I remember my mom was thrilled. Over the phone to the salesman, she said, "We'll take it!" We all had great times in that car. We went car camping along the Skyline Drive in Virginia. We drove to the beach in Delaware. My dad piloted slowly around our neighborhood streets as my brother and I delivered the Sunday *Washington Post*, hopping off the car's tailgate with a few heavy papers under each arm, then hopping back on.

Our happiest times as a family were in a station wagon that was white with a red interior. Now it's not so surprising that my sister had copied me when she bought her car. Or did I copy her? No, neither one of us had been in touch with the other about these purchases on opposite sides of the country. We were just attracted to these vehicles, which resembled the one we'd had good times in. It doesn't say anything about genes or about fate. It does tell me that what we commonly think of as free will is influenced strongly by life experiences already in the bank. We are never really free of our past.

As you (I hope) and I work to become agents of change, it helps to be aware of those experiences and their lingering effects on us. My parents were both college-educated, and they both ended up serving the United States in World War II. My dad had a pretty rough experience in that Japanese prison camp. My mom worked in an underground office as an officer in the Navy. Both became progressives in their politics. They didn't talk about it a lot, but I'm pretty sure they both saw the war as a tragic waste of humankind's intellect and treasure. At the same time, they both appreciated the enormous role government could play, not only in conducting a war and repelling an enemy but also in providing a decent quality of life for its citizens. Those values rubbed off on me, too, in ways that are a lot more meaningful than my lingering fondness for a sharp white car with red upholstery. If I turned into a raving progressive—and I'm pretty sure I did—Ned and Jacquie had something to do with that. I know where my values came from, and I feel good about the source.

As for my brother and sister, they diverged over the years. My sister went to college in Danville, Virginia, married her college sweetheart, and raised three kids. She worked for the city of Danville in various capacities, including as a 911 dispatcher. Her family lives close to the border with North Carolina in a quite conservative part of the country. My sister is generally liberal like me, but her kids are not exactly inclined to my style of politics. Although we share a great many gene sequences, they see the world from an almost completely different point of view. They are more than a little resentful of government and its intrusions. Meanwhile, my brother remained in the Washington, DC, area. He raised his family there, and they all share the progressive points of view. Spelling it out: Familial genes matter, but one's environment and peer influence matter a great deal, too. Nature and nurture are still not destiny, though.

It's like the white four-door hatchbacks with the red interiors. Our free wills are influenced by the sum of our experiences—family, friends, the social environment, and, increasingly, the online environment, as well. My sister's kids and I got nurtured differently, and even though our inner natures are largely identical, we ended up with quite different political views. My brother's kids were raised in the big city, exposed to people and experiences similar to the ones in my life, and ended up embracing a view of the world similar to mine. Within this diversity, interestingly, we still all laugh at the same hilariously funny jokes. (I'd share one of these jokes, but then you'd realize that my jokes are not especially funny. That might affect how you judge my siblings, who are no longer responsible for my humor choices.)

In my family, just as with my dad and his buddies in the prisoner-of-war camp, a shared sense of humor overrides just about anything else; politics-schmolitics. It's reasonable to me that our sense of irony is a product of the way we perceive the world, especially the way we perceive and interpret the actions of others. Perhaps part of what keeps a family together is what makes them laugh, and for that maybe there is no free will. It strikes me as a worthy hypothesis, but testing it would be serious business—I mean it would be no joke. (Uh . . . sorry.)

A great paradox is hidden in this whole discussion of behavior and free will: The researchers who study it (and people like me who follow along) are trying to figure out how our brains work, but the only tool we have to do that figuring is . . . our brains. Brains trying to understand brains.

It's a subset of the much broader problem of the inherent subjectivity of the human mind. Scientific method was developed to get around that problem by codifying techniques that account for and work around our flawed intuition as best we can. The best chance we have of ensuring that we are not blindly ruled by the strong influences of our family, friends, and social conditioning is to follow that method: Observe, hypothesize, experiment, compare result with expectations, and—most importantly—start over. It's a way to escape the echo chamber in your own head. It's a workout routine for exercising your free will.

The scientific method forces us to question our assumptions and weed out the ideas that are based on gossip, opinions, past conditioning, and all the other baggage we carry with us. That approach is closely akin to one of the guiding questions of investigative journalism: What do I know, and how do I know it? Unfortunately, the human brain has a built-in mechanism that directs our behavior the opposite way. It is known as confirmation bias, the tendency to confirm our own assumptions as valid and true.

For a scientist conducting an experiment, confirmation bias is a big, big deal. In this sense, there clearly are limits to our free will. Our unconscious mind can lead us to see the answers that we are conditioned to expect, rather than to find the real ones we are looking for. The consequences can be severe. Medical researchers who were convinced that widespread screening would reduce the rate of breast cancer deaths found that result even when later studies could not verify it. It's likely that large numbers of women had needless surgery as a result. Such biases crop up whenever we approach a problem knowing what we want the answer to be. If your boss (or child, spouse, friend, etc.) has ever asked for advice but then obviously listened only for the comments that matched what he or she already wanted to do, you know what I'm talking about.

To overcome your own confirmation bias, you can draw on the ways that researchers deal with it. We go about our work in the opposite way of what many people imagine. Instead of designing experiments to prove a hypothesis right, we work hard to develop schemes or techniques that would show our hypothesis to be *wrong*. We actively work to set our brains free of their biases. When humans codified the process of scientific inquiry, it was a major leap forward in thought.

The way I see it, all life exists on a gradation. We are on the high end of intelligence (smarter than most dogs, at least, though maybe not smarter than *your* dog), and I submit we are on the high end of intuition and the ability to reason. The scientific method—the nerd way of thinking, I might say if I'm feeling a little self-congratulatory—is our attempt as a species to stretch beyond our stubborn evolutionary limits. It is our most meaningful appropriation of free will, or at least as much free will as our brains allow.

We seek theories that can make predictions about the cosmos, our planet, and ourselves. We want to know the laws of nature. Is our curiosity itself part of our evolutionary programming? I think it has to be. Our ancestors who were not curious about the world around them got outcompeted by the other nearly naked gals and guys who sought answers to life's problems. Where do I sleep tonight? What will I eat next season? You could think of this as an early impulse to rebel against the idea of destiny and take the future into our own hands. If we know what's coming, we can prepare and change the outcome (for example, not starve to death next winter). By extension, maybe our desire to choose freely is programmed by evolution, too. You can go crazy thinking about these things, albeit while having fun. All the evidence tells me that humans, more than any other species, have transcended their evolutionary impulses and get to choose how they want to behave toward each other and toward planet Earth.

I believe we each have a gradation of choice-making ability within ourselves. I bought a car for reasons I couldn't specify and probably didn't consciously know. I spent several months trying to decide whether to quit my day job as an engineer to focus on writing and performing for a living. I took at least as long to come around to the idea that I, with virtually no management experience, could manage a multimillion-dollar nonprofit organization, The Planetary Society. The bigger the decision, the more I activated the nerd side of my brain, and the more I recognized the need to grab information in everything-all-at-once style and then filter it like crazy. When I finally reached the moment of "Well, let's go!" I believe the process that brought me there was, in a very significant way, free will.

What I am absolutely certain about is that we will never attempt great deeds and never—dare I say it?—change the world if we do not embrace our human freedom. On the most basic level, that means getting past the easy cynicism of grumbling about a rigged system or feeling like no one person can make a difference. We have to learn to see the problems around us clearly, without the fog of confirmation bias. We need to make nerd honesty a way of daily living, not a specialized tool for scientific investigation.

Then we can really begin to figure out what's going on and which solutions will do the most to fix our problems. The deliberate method of rational investigation—approaching, addressing, and solving those problems—is a transcendent act of free will, and we have to choose to exercise it. An even greater expression of freedom, though, is taking the next step and putting the scientific method to work. Letting it guide your life so that, at every moment, you see the greater possibilities around you. Realizing your personal potential. Recognizing your responsibilities. Turning back global warming. Reducing poverty. Expanding access to information. We are never more human than when we reject the idea of an imposed destiny and use science to change the world in big, bold, and very free ways.

CHAPTER 19
Time for Measured Urgency

It's currently fashionable for people to brag about their extraordinary skills at multitasking. As you know, I've never (okay, seldom) been a follower of fashion, so let me honestly and unfashionably say that I consider multitasking to be a modern hoax. The title of this book might mislead you into thinking that I recommend starting countless jobs at the same time. Well, that's not what I mean—not at all. It's why I put so much emphasis on filtering. I want us all to use good judgment not just about what needs to be done but also about when to do it. In other words, if you want to be an effective agent of change, you also need to filter your sense of urgency.

I'll start with a few huge examples. We have to address the needs of people in extreme poverty, and we need to prepare for the rapid changes in our seacoasts and agriculture brought on by rising global temperatures. These challenges are urgent, but they are going to require our attention for decades. We also need to search for life in the solar system and beyond. That's not nearly as urgent, but to me it is very important; it, too, is a decades-long undertaking. To address these projects, we need to take in everything all at once, sorting through all the best information and designing well-considered plans. Being attentive to priorities is a critical part of that process, and recognizing the fallacy of multitasking is an important first step.

Inattention is a chronic problem these days. Where I live in Los Angeles, I've seen countless car wrecks that could have happened only if someone was simply not watching the road. Should you ever be driving behind me in slow traffic, you might notice the custom license plate frame I got in response. It proudly reads, "TRY MONO-TASKING." So far as I know, "monotasking" is not in a dictionary—not yet. But it may be soon, because no one really multitasks; no one does two things (or ten things) simultaneously. Even the circus performers who juggle while riding a unicycle are doing just one thing: performing, with their subconscious running the show. They're not thinking about two tasks at once. If one of those performers were genuinely multitasking—using the reasoning part, or conscious part, of her or his brain to control each individual component of the act—there's no way she or he would get anything done at all. She'd be inattentive to the real goal, probably ending up flat on her face or back on her tailbone.

Instead of multitasking, it looks to me that what the most successful people are doing is managing a great many things in parallel, paying full attention to each individual task as immediate priorities demand. This style of sequential monotasking reminds me of another guy at the circus, the one whose performance includes the wonderful spinning plates. Traditionally, Khachaturian's "Sabre Dance" plays in the background while he works. He spins one plate on a flexible dowel. Then he spins a second, then a third. Before spinning a fourth plate on a fourth dowel, he dashes back to the first plate and revs it up a little to keep it going. Same with the second and third plates. Only then can he dash back to the fourth plate, and so on. Instead of multitasking, he's just *tasking*, spinning one plate after another while paying close attention to which one needs another boost. I feel this is how we get something, or really anything, done. No matter how complex the action, no matter how fast or slow, we tackle the necessary steps in the necessary sequence. It's pretty nerdy. When people forget to act sequentially, you get those freeway car crashes. Maybe someone tried to be equally attentive to the road and the radio at the same time, or decided that reading a text message was more urgent than assessing the distance to the car in front.

Multitasking is the formula for a car crash, whether literal or figurative. But what I'm talking about—whether you call it monotasking, or just tasking—allows you to get things done for real. But it still doesn't tell you what tasks you need to focus on. It doesn't tell you when it's okay to look away from the road for a moment, or which plate needs another spin *right now* if you want the act to keep going.

This all reminds me of the days when I was writing comedy full-time on the *Almost Live!* show in Seattle. I came up with a sketch called "Measured Response—or Overreaction." It starts with coworkers sitting around a lunch counter. One guy knocks a cup of hot coffee over, spilling it into the next guy's crotch. The victim calmly pours a glass of ice water into his lap; a large cloud of steam (from a fog machine) puffs up, showing us just how hot that coffee must have been. The announcer explains that we just saw a "measured response." Next, a woman tells a man that she has to cancel their dinner date. In response, he smashes a bottle on his head and starts chewing a couch cushion. The announcer explains that it was an "overreaction." Finally, a man driving a large car notices a parking space, but just as he is backing up to pull into the space, a much smaller car whips in and takes it. As the small-car driver gets out of his car, the first guy starts a fistfight, punching wildly. The announcer explains that the stolen parking space was an offense worthy of strong retribution and states in a deadpan voice, "measured response."

The sketch is funny (or so it was intended) because, like all comedy, it is based on an essential truth about human nature. We are constantly evaluating which circumstances require just a little effort on our part and which ones demand that we go all-in. Measured response applies to every type of project, from making a sandwich to building a citywide water-sanitation system in the developing world. It applies to the choices between space exploration and antipoverty programs and to the multitude of individual actions they each encompass.

We each get a painfully limited amount of time on this planet, which we can fill with only so many tasks. It's necessary to ignore a great deal of what goes on around us, but it is also necessary to pay close attention to the things that matter most. We nerds, who are especially attuned to the big picture and the big issues, have a special responsibility to address and resolve the troubles we see. So how do we reconcile these conflicting demands? We need to become measured-response experts. We just need to measure a little more carefully than that big-car driver in the sketch, the one who rolled past a perfectly good parking spot.

I often hear climate-change deniers (perhaps better called "climate-change whiners") complain that there is no immediate solution to any problem inherent in the world's warming. Any idea we came up with would just take too much time, they argue. From this they conclude that, even if the scientists' warning is correct, there's no point in taking drastic action because it's too late to do much of anything about it anyway. My first thought is usually along the lines of "How do you know it will take too much time if you also think that global warming is not really happening?" But once I get over myself, I have to admit that there is an important question within their impotent whine. It's another problem of measured response.

We do need to know the pace of change. This question of speed, and the associated one of how much urgency is called for, is the focus of intensive research by climate scientists. They track the melting of glaciers and the rising of sea levels, data which will tell us how quickly we need to prepare for coastal flooding (I'll present much more information about that in Chapter 24). They monitor Earth for temperature changes, then test the observations against their detailed climate models. They research the record of climate change over the past hundreds of thousands of years. All these actions draw on an ancient principle: If we want to change the world, we need to understand the processes of the world and the rate at which they unfold.

Measured response is not a new concept. It is as old as the desire to be in control of nature. It goes back at least to the invention of the first calendars, many thousands of years ago. I maintain that the calendar may well have been the greatest invention in history, greater in many ways than the invention of the wheel. There were pre-Columbian cultures in Central and South America that had no wheeled carts or vehicles, but they had exquisite systems for reckoning time. The calendar was literally a matter of life and death. It was created to ensure the proper level of urgency at any time of year. For example, our ancestors used calendars to know when to plant crops, when to prepare for heavy seasonal rains and flooding, when to stockpile wood for winter fires. People had to keep their priorities straight throughout the year.

For many centuries, horological nerds worked and worked to refine their reckoning of time. When we buy a calendar from a regular store, we presume that the days and months will be properly accounted for and arranged, but it took thousands of years to develop the most basic tool of measured urgency. There's another intriguing, oft-overlooked aspect of this early history of the calendar: the role of religion and mythology. A calendar that was accurate to within a few days might have been sufficient for planting crops, but if your religious ceremonies required you to know exactly when to expect a lunar eclipse or an unusual conjunction of planets, you needed much greater precision. You had to recognize long-term patterns of celestial motions and anticipate how the various objects would move far into the future. In response, priests and shamans studied the sky in detail and learned to predict the exact motions of celestial objects. They invented astrology, the ancestor of today's astronomy, and achieved a remarkably powerful tool for measured urgency. (Just to be clear, though, knowledge that was remarkable 4,000 years ago is now absurdly outdated. Nerds have, uh, made a little progress since then. Would you trust your life to ancient Sumerian medicine, for example? How about good old barber-pole bloodletting when you're really sick? I hope you see what I mean.)

The challenge today is another form of information overload. We know so much about so many different problems, and have so many different measures of urgency, that it can become completely over-whelming. Even if you take my word—and you should, of course—that multitasking is a hoax, it's easy to get paralyzed by competing priorities. You cannot attack every task with the same amount of effort every-where, all the time; you would go crazy (in my case, crazier perhaps). If you treat every problem as an immediate number-one crisis, you can easily get sucked right back into the black hole of multitasking. That holds true not just for people but for our society as a whole. Perhaps our public officials routinely take too small a temporal perspective and completely lose focus. Or maybe they're doing our bidding, and it's us, the voters, who are to blame. And indeed, all of us rebound to our short-time point of view once in a while, and take too big of a temporal perspective. We think everything will pass into the past, so we take no action. Either way, the result is . . . inaction.

Failure at measured response has very real human consequences. Every year, hundreds of millions of people suffer from dysentery, malaria, and other controllable diseases because—wow, where would you even start? You surely know the feeling yourself. If you ever donate to a charitable cause, your mailbox and email box are soon clogged with solicitations for more donations. Twitter is a riot of outrage and calls to arms. Your Facebook feed is probably clogged with requests to join this group, march with that group, or give money to a hundred different worthy-looking causes. If you overreact to the first ones you see, you may end up misapplying your energies on a marginal project. If you underreact, you may end up missing out on something truly effective and meaningful. And forget about trying to engage with all of them; there aren't enough hours in your life, let alone enough hours in the day. Meanwhile you (or maybe not you; you're far smarter and more sensible, but someone like you) end up in an online shouting match about science or politics that sucks up all your energy. It's an unproductive overreaction, like the jilted lover chewing madly on the sofa cushion.

What we need, then, is not just measured response but measured urgency. This is where nerds need to activate the second kind of filtering that I alluded to at the beginning of the chapter: temporal filtering. It is the way you figure out the correct measure for your measured response and the best order for your sequential tasks. Mastering the way our society regards time has to be at the core of the everything-all-at-once approach.

And how do you do that? Here I would like to bring you back to my beloved upside-down pyramid of design. The pyramid is more than a guide to how much things can cost during the design process. It is also a timeline that runs from its pointed base to the wide crown at the top. This timeline is a complex one that does not strictly move in a line. In the car business, for example, manufacturers have to start cultivating customers before they finish building vehicles or troubleshooting computer code, so the different levels of the pyramid can overlap in time. There's a wonderful word in mathematics that describes the way time is mapped through the pyramid of design: "orthogonal," meaning "at a right angle to." The passage of time is orthogonal to the pyramid. Time affects every step in the ziggurat, but it is not bound to any one step.

Think of the pyramid as a visual depiction of what I mean by monotasking rather than multitasking. We use critical thinking to filter information, which shows us which problems need solving. Filtered information also guides us on how to solve them. The upside-down pyramid of design helps us filter out *when* to solve them. You can treat even a huge problem like climate change as a matter of design and break apart the different components: upgrading the electrical grid, regulating carbon dioxide emissions, building more wind turbines and installing more photovoltaics, building seawalls, expanding reservoirs, etc. We cannot shut down every fossil fuel–powered plant tomorrow afternoon, for example, because it would bring our economy to a standstill before we got the renewable electricity infrastructure built. Each of those components, in turn, can be broken down into its own pyramid with its own scientific and technological priorities—its own list of measured urgencies.

The principle applies to other pressing issues. We cannot confine all unvaccinated people to quarantine at their homes until they are vaccinated. Instead, we have to manage the problem from the bottom up and with the orthogonal flow of time. The same holds true for your own individual contributions to the great solutions of the world's great problems. I admit that I, too, sometimes get whelmed (somewhat short of overwhelmed) by all the tasks ahead. But then I recall the pyramid of design, shift my perspective, think of it as a stack of tasks, and assign urgency to move from the design vertex to the finished product at the top.

When it comes to the task of procurement, be it groceries or laboratory glassware, I am continually making lists of items to shop for. It is a small and eminently manageable problem. Within each list is an implied sense of priority. There are groceries that must be purchased today, or I'll have nothing to eat for dinner. There are grocery store items that aren't edible, like paper towels and laundry detergent. Then, there are items to be acquired from the hardware store or online—important but less urgent. Each item is nested or placed according to not only where it will be acquired but when, both when I'll find it as I route myself through the store and when I may really need it. It's a small exercise that simplifies my shopping and ultimately reminds me of how a nerd like me accomplishes anything. I do all this without thinking, and I'm sure you do, too. You already have an instinct for everyday measured urgency. All you have to do is learn to scale it up and apply it to tasks that you probably don't think about that way—yet.

The key to the pyramid and the orthogonal time coordinates is that we nerds can break any task, tiny or huge, into what I might call tasklettes. In the context of the auto industry, I summarized the four levels of the upside-down pyramid as design, procurement, construction, marketing. For evaluating measured response and measured urgency, we need to reinterpret the levels more explicitly in terms of time.

- Identify the problem (e.g., we need water for a village or valley).
- Design a solution (e.g., we need a properly placed dam).
- Procure the support and the resources (e.g., get the local government and the villagers to embrace the idea and buy the concrete, rebar, and penstock actuators).
- Get it done (e.g., build the dam).

This list is very much like the upside-down design pyramid. It starts with nerds seeing the problem in a scientific or technical way, and it often means doing all four of these things in overlapping sequence, like a world-changing plate spinner. In order to build the dam, you have to have support of the people. You also need the people to see a dam as a solution to their water and agricultural needs. And ultimately, you sure as hell need the people to build the damn dam thing. You can always start by evaluating whether the action you want to take, or the cause you want to support, has cleared the first two levels. Does it address a specific problem? Does it describe a specific solution? If so, it can move up the hierarchy of urgency. (Arguing with strangers on social media is never, ever going to get past these filters. If you have that inclination, do something else immediately.) For any project that aims to address the higher levels, see if it has already completed the first two. If not, you are dealing with an overreaction, a premature attempt to build results without putting in the difficult foundation work.

Fundamentally, we have to embrace the idea that all our tasks, even the most enormous ones—addressing climate change, providing clean water, supplying reliable renewable electricity, and opening access to electronic information for all—can get done by breaking them down into manageable taskettes and filtering them orthogonally, by urgency. That is the way that good impulses turn into workable policies and life-changing projects. Nerds have to lead, to show the world what we can accomplish when we set the right goals, equip ourselves with the right information, and act on the right measured rhythm. And if we make some mistakes along the way, well, nerds have a great system for dealing with those, as well. Read on!

CHAPTER 20

A Mind Is a Wonderful Thing to Change

In 1987, Carl Sagan was addressing a meeting of the Committee for Skeptical Inquiry when he said something that has stuck with me over the years: "In science it often happens that scientists say, 'You know that's a really good argument; my position is mistaken,' and then they would actually change their minds and you never hear that old view from them again. They really do it. It doesn't happen as often as it should, because scientists are human and change is sometimes painful. But it happens every day. I cannot recall the last time something like that happened in politics or religion."

Changing your mind. What a powerful idea! The possibility of change is essential to maintaining an honest, open view of reality. If you truly want to apply the nerd standards and find the best solutions, you have to be ready to give up a flawed position in response to new evidence. That principle is so simple to praise yet so difficult to practice.

Whether because of pride, ego, or some hard-wired ancient instinct, we connect changing opinions with being wrong and thereby being weak. But here's the thing: If you want to be right all the time, sometimes you're going to have to update your views. You know it. I definitely know it, because I recently changed my mind about a big scientific idea—and I did it in a very public way.

In the first edition of my 2014 book *Undeniable: Evolution and the Science of Creation*, I have a chapter about GMOs. At the time, my attitude toward them was not very positive, though not for the most common reasons. I had no reason to doubt that modified foods were and are safe to eat. Researchers have fed engineered corn and soybeans to lab rats and monitored their well-being. Not to creep you out, but researchers perform careful autopsies to see if the animals suffered from unusual rates of tumors or other medical anomalies. Those tests have revealed no evidence of harm from eating genetically modified foods—none. So I've long been confident that people won't suffer any harm, either. No, I was concerned about the unpredictable effect of GMOs on ecosystems.

My skepticism was rooted in ideas I've been thinking about since that first Earth Day in 1970, when I became aware of how easily humans can upset the balance of nature. This was the era of Rachel Carson's *Silent Spring* and *The Sea Around Us*. Rachel Carson was a very influential marine biologist; these two books were bestsellers and are still discussed today. Whatever you or historians may think of Carson in retrospect, in the 1960s she had an important cautionary message about poisoning our environment. I took her deeply cautionary message seriously. Since we cannot know exactly how our actions are going to affect the environment, I reasoned, we need to be extremely careful about introducing any major change. That caution certainly seemed to apply to the widespread planting of genetically modified crops. I was worried about unintended consequences.

I conjured a scenario in which scientists engineered a plant that is resistant to a certain pest caterpillar, for example. The modified version of the plant might make a protein that the caterpillar finds toxic; this is very close to how a real crop called Bt corn works. (It's named for *Bacillus thuringiensis*.) As a result, the caterpillar and butterfly population goes way down around the farmland where people are planting the new, resistant crop. All well and good, except what if there is a bat that relies on those butterflies for food along its nightly flight path? When the butterflies go away, the bats change their behavior. Maybe some of them starve. Either way, they no longer reach their preferred destination: a nearby lake where they normally eat vast numbers of mosquitoes. Then the mosquito population at the lake goes wild, and those mosquitoes then spread diseases to humans and animals all around the region.

It's not an unrealistic scenario. These kinds of ecological domino effects happen all the time in nature. My argument was that you cannot know any organism well enough to anticipate the consequences. It gave me pause.

I was also struck by the compelling idea that we already have enough food to feed everyone in the world, even without genetically modified (GM) crops. The problem is food distribution rather than supply. Meanwhile, I was watching all this public controversy over genetically modified foods. A lot of people don't want to eat them. Some farmers were upset about intellectual property rights—about who owns seeds that are uniquely engineered to fight thrips or corn borers or caterpillars or some other pests. So I reasoned, maybe we don't need GM foods and don't need to spend government dollars issuing patents and defending the legal rights of corporations that want to create them. I deliberated over these points with my esteemed book editor Corey Powell, thought through my argument, and wrote all about it in *Undeniable*.

Then some things happened that made me start to rethink my position. In December 2014, I went to a public debate about genetically modified food in New York City. It was just a few weeks after my book had come out, and I was surprised by what I heard. On the skeptical, anti-GMO side was Margaret (Mardi) Mellon from the Center for Food Safety. I had met Mardi a few years before, when she was the agriculture policy director at the Union of Concerned Scientists. Her argument centered on sustainability: GM crops, she claimed, have not cut the use of insecticide and have only increased the use of herbicides. Mardi raised some interesting questions, to be sure. But my impression was that her arguments were overwhelmed by the responses from Robb Fraley, the chief technology officer at Monsanto. He's one of the guys who invented plants engineered to tolerate the herbicide glyphosate, and I was impressed both by his technical expertise and his obvious environmental concern. Afterward I spoke further with both Mardi and Robb. I wanted to process all this new information.

Robb invited me to Monsanto to see what they do there. I paid my own way and headed to St. Louis. I put a lot of questions directly to the Monsanto researchers, and I found their answers highly persuasive. I tried my best to put emotions and assumptions aside and to gather as much high-quality information as I could about the true potential of genetically modified foods. Along the way, I learned some surprising things. Using modern gene-sequencing machines, they can know every gene in a plant's DNA within about 10 minutes. They can assay both the natural and the engineered genes and then determine exactly how those genes will behave and interact with other organisms (like insect pests) once the plant is out in the field.

Robb showed me how careful the Monsanto agronomists are when they raise their modified test crops—first in greenhouses, then in controlled refuges—and the discipline with which they plant and monitor those test crops. I found the level of specificity remarkable and satisfying; I realized that there was far less uncertainty in the process than I had believed. Robb and the staff of scientists at Monsanto also made the compelling point that *all* farming is unnatural. Farms would not exist without humans pushing around plows and tills and, more recently, spraying herbicides and pesticides. We've been altering crops through careful breeding for thousands of years. There are no "natural" farms and there are no "wild" crops. We can plant and harvest GMO crops, or we can plant and harvest other crops that we've created using traditional techniques, without these high-tech modifications. Either way, our farms are—from nature's point of view (if she has one)—inherently artificial.

I ended up rewriting that chapter in *Undeniable* for the paperback edition. In the new version, I concluded that my concerns about the ecosystem were a manageable problem and that genetically modified foods could be an important part of the future food supply. I kept myself honestly open to new information and . . . I changed my mind.

I achieved a fuller understanding of what the modification techniques actually do. The scientists at Monsanto and the other companies and labs have found ways to incorporate genes that enable plants to create proteins that are normally found in soil bacteria. One of these proteins targets a common pest, the larval form (caterpillar) of the European corn borer. If the caterpillar eats the modified corn, that protein accumulates and crystalizes in the caterpillar's stomach. The pest dies. Corn borers never eat soil bacteria, but after the scientists are done, the borers cannot eat the corn, either. The corn has been modified to make the same deadly-to-caterpillars protein. Since the caterpillars can no longer eat the corn plants, they go looking someplace else for food and leave the farm field alone. That's good.

Still, killer proteins sound a bit scary. Here's an important thing to know. Organic farmers spray their crop plants with a liquid form of the exact same protein, *Bacillus thuringiensis* (that's Bt) toxin. It's an organic pesticide in the sense that it is made by a naturally occurring organism. The only difference between what the organic farmer and GMO farmer are doing is that the former is spraying the chemical onto the plants rather than having the plants make it themselves. The modified plant is better at resisting the corn borer, and it requires less spraying. The protein gets into every part of the plant, but it does not affect us. The corn borers, on the other hand, are more than a little bothered by it; it kills them. Look, I'm over 60 years old. I've eaten popcorn every day, sometimes twice a day, for most of the past 2 decades that corn has been genetically modified. I can't be sure when you might be reading this book, but it's likely that I popped myself some popcorn last night. I'm not concerned about the GMOs, and I'm comforted by the idea that the kernels were exposed to the lower amount of chemicals.

Another extremely successful genetic modification creates crops that are impervious to glyphosate, a serious-business herbicide. Glyphosate is better known by its brand name, Roundup, a name that people love to hate. Roundup is amazingly effective when it comes to killing plants. When it was invented, two things happened: 1) Everybody was afraid of it, and 2) everybody started using it. Farmers, gardeners, homeowners, everybody uses Roundup because it is so good at killing weeds. But if you put the right gene in your crops—soybeans, cotton, corn—you can use Roundup to kill the weeds without hurting the crops. The resistance gene was first isolated in petunias, of all organisms. It's amazing. Farmers who plant these "Roundup Ready" crops do a good bit of spraying to get rid of the surrounding weeds. But without Roundup Ready crops they'd be spraying some herbicide anyway, and they'd be doing a lot more tilling: turning the weeds over to kill them by exposing their roots to sunshine. Farmers need to control weeds somehow, and glyphosate has proven to be a boon.

Scientists who work on genetically modified crops claim that spraying Roundup is actually better for the environment than traditional weed-control methods. Their explanation is that tilling dries out the soil and kills off a lot of the natural soil ecosystem. They also assert that, unlike traditional herbicides, glyphosate breaks down quickly in the environment. It completely disappears after a few weeks. I try hard not to accept bold statements like that without some verification, so I took a look at the peer-reviewed literature. I've read papers and I'm satisfied that it's true. The genetic diversity of microbes and multicellular soil species (worms, insects, larval insects) is much higher in Roundup-treated crop fields than in tilled fields. And glyphosate compounds are a type of salt that does indeed chemically decompose. In that important sense, at least, the Roundup fields are healthier than untreated, organic, tilled fields. That's a consequence I hadn't anticipated, in a good way. These features of GMO farming also helped change my mind.

Even with all this encouraging information, I still had a couple of concerns gnawing (corn-boring?) at me. In my earlier hypothetical scenario, I described a hypothetical species of bats that wouldn't have enough butterflies to eat because of the introduction of GM crops. Something kind of like the first step of that story really has been happening. In spraying their modified crops with Roundup, farmers have been killing off a lot of milkweed plants, common weeds that grow on and around farms. Monarch butterflies lay their eggs on milkweeds, and when the larvae hatch, they feed on milkweed leaves—and milkweed only. It turns out that milkweed cannot tolerate even a little bit of glyphosate. So, farmers spray their fields; milkweed goes away; monarchs lose their habitat. Roundup Ready plants are part of the reason why the monarch population has dropped by up to 80 percent in recent years. It's a textbook example of unintended consequences.

But there's good news. The monarch population is bouncing back. At an April 2015 meeting in Minneapolis called the "Monarch Venture," a group of ecologists, activists, and businesspeople got together to address the plight of the monarchs. I went there and listened all day. What was most exciting was that everyone attending the event seemed to be there to listen, too. The hippies and the corporate pigs talked to one another, like open-minded nerds, and they found common ground. They got together and created a plan based on the best science they had. They agreed that farmers would leave refuges of milkweed along monarch "flyways," natural highways in the sky where the wind blows south to north, aiding monarchs and birds in their annual migrations. The "Venturers" space the milkweed refuges close enough to one another to sustain migrating monarchs. Last year, the monarch butterfly population wintering in Mexico was nearly four times what it had been the year before. Good weather undoubtedly helped, but it's encouraging nevertheless.

At the end of all my investigations, I also did an internal reality check. Was I swayed by all the attention I got from the folks at Monsanto? I really don't think so. I watched carefully. Everything I saw at Monsanto, and everything I learned doing a lot of additional research, tells me that the people there genuinely are trying to help farmers improve their yield. Of course they are trying to make money in the process. So are farmers. So is the guy at the organic fruit stand.

Financial incentive doesn't tell you who is right and who is wrong. The meaningful way to approach the issue is broadly, openly, honestly, with everything all at once. And *that* is how I changed my mind.

» » »

When I wrote my new chapter and announced that I had rethought my position on genetically modified crops, the scientific community was excited. I still get emails about it. A few researchers were specifically pleased to have me as an ally in promoting the benefits of GM foods. Many more wrote to tell me how cool they thought it was that Bill Nye changed his mind based on a reconsideration of the scientific evidence. They loved seeing that change is really possible.

I found it challenging to abandon my earlier ideas about GMOs, but I managed to maintain an open mind through the whole process. I talked to many scientists and read through the literature, trying hard to avoid selecting the evidence for any particular point of view. I flew to St. Louis and Minneapolis, and I did those trips on my own dime. It took time and effort, and also some soul-searching. I had to make peace with the unsettling thought that I had been wrong. I also had to block out a lot of information that was clearly biased. Challenging your intuition is hard, but it is harder still when there are a lot of people on your (current) side urging you not to.

I recently waded into an anti-GMO rally in Manhattan while a documentary crew was filming the scene. I was struck, and not in the good way, by the ignorant views of some of the protesters. I tried to have a conversation about genetic modification with a few of the demonstrators, but the people I spoke with quickly derailed the discussion, or they conflated GMO science and economics with a bunch of other political issues. They did not seem open to listening or to changing their opinions—ever. In my view, that hard, ideological attitude made the protestors look silly. That in itself doesn't tell you anything about whether genetically modified foods are good or bad, but it does tell you about the difficulty of nerd honesty. Many of the people at the rally thought they were being scientific. They had certain statistics or examples or arguments that they liked to recite. But they were lacking the perspective that is crucial in science. They weren't filtering information from the Internet; they were uncritically blocking anything that might cast doubt on their side of the argument.

Changing your mind is not a one-off action. It emerges from a habit of mind that you embrace as part of your life philosophy. When you are trying out a new idea, you have to be ready to say, "It looks like this thing is working but I'm not sure." When you weigh in on a controversial topic, you have to be ready to admit, "I've always thought that thing was true, but my own reasons aren't enough. I need to get more information, even though that will cost me time and effort." You have to internalize a restless curiosity. Hold on to a childlike sense of wonder so that "prove it" is not an angry challenge but an exciting opening to discovery. All this has to become the normal way you think. Your priority has to be wanting the most accurate facts and the best possible data, rather than being right about your hypothesis. My GMO investigation reminded me of this fundamental idea—very strongly.

You also have to be aware of our human tendency toward confirmation bias that we looked into back in Chapter 18. It is something I talk about in my public lectures, too. We did a whole *Bill Nye Saves the World* show on it. Let's review: Say you grow up with a parent who believes in astrology. You'd tend to grow up believing in astrology, as well. When someone shows you evidence that astrology is bunk, you'd probably be troubled. "I spent my whole life reading my horoscope, and now you're telling me it's not true?" Yes, that's what we're telling you. You'd very likely look for counterexamples. You'd say something like,

"Three days in a row, my horoscope said something that applied to me perfectly." That's confirmation bias: faulty filtering that lets in supposed facts that support what you already believe but block out everything else. It takes a long time to overcome it. That's why I mention it again here. In my experience as an educator and just as a regular citizen, people have to be exposed to skeptical thinking many times before they are ready to question their assumptions. You also have to believe in your own fallibility. Most of us think we're above-average drivers, right? If only. On average, we are average drivers, and average thinkers, too. How could it be any other way?

Even in science and engineering, there is a large and continuous problem of confirmation bias that we must continually push back against. It's wired into our brains to see what we expect to see and to trust our own opinions. Physicists have to be careful to make sure they are seeing a real particle and not just a signal that they expect to see, ostensibly accounted for by their pet theory (pet particle?). Cancer researchers work very hard to make sure they are seeing a real response to a new therapy, even if they have a lot of hope and professional pride riding on a positive result. Climate researchers worked for decades to make absolutely sure that the effects they were detecting were real. It's a difficult business. Even after years and years of thinking about science, I guarantee you there are things I believe that are absolutely wrong (but I also guarantee you that the seriousness of climate change ain't one of 'em; there's just too much evidence).

The good news is that, with humility and an open mind, it is always possible to learn more. You can arrive at a more honest and accurate understanding. As much as that open nerd-style may be contrary to your deep impulses, it taps into one of the best aspects of human nature: the ability to accumulate knowledge and improve the way we live.

If you keep thinking about things the same way you always did, and looking at the same sources who frame the issue the same old way, you will keep coming back to the same answers and the same assumptions. Embracing the possibility of changing your mind requires being willing to accept different inputs. I keep talking about how difficult that is, but in the spirit of change, I'm going to think about change itself in a new way. Why do people love going on vacation? They want to escape from their routines. They want to see a new part of the world. They want to get better at a sport. They want to see people they haven't seen in a long time, meet with old relatives, make new friends. That's exactly what nerd openness is all about. It means that, at a moment's notice, you are always ready to pack your bags and take a vacation from the old ways you are currently thinking about the world.

I love—well, let's go with *like*—getting out of my comfort zone. It's important to get out of your bubble and see what's going on in the bubble next door. That doesn't require long journeys and advance planning; it's something you can do constantly in small ways. I am not a Fox News kind of guy, but I watch Fox News because I need to know what other people are thinking and hearing. If you are convinced that the other side is not listening to you, that is all the more reason to listen to them and honestly evaluate what they have to say. That is the only reliable way to change your mind, to have a chance to change someone else's mind, and ultimately to have a realistic shot at changing the world.

PART III

How to Change the World

CHAPTER 21

Are You an Imposter?

When I took on the job as CEO of The Planetary Society in 2010, I didn't believe I was up to the task. My predecessor, Lou Friedman, cofounded the society and had been running it for 30 years. Meanwhile, I knew almost nothing about management (not the same as leadership) and nothing about the structure of nonprofits. The Planetary Society had employees on a payroll with sick leave and insurance benefits—and I was suddenly in charge of all of them. I felt pangs of imposter syndrome, an irrational fear that derives from expecting too much of yourself and from assuming that other people must be more capable than you are. It's a distracting psychological bug in your brain that holds you back from doing the things you want to do. It is, in essence, overly critical self-critical thinking. The nerdier you are, the more susceptible you are to it. Since you are reading this book, there's a good chance that you've experienced some form of it. And I hope we all learn how to deal with it.

Part of what changed my self-perspective was my interactions with three people. One is Neil deGrasse Tyson, who sat on the board of directors of The Planetary Society; you've probably heard of him or watched him hosting the new version of *Cosmos*, or maybe you've listened to a few of his (and occasionally my) *StarTalk* podcasts. Another influential guy for me is Dan Geraci, the chairman of the society's board. Dan has an investment firm based mostly in Canada, and he has tremendous experience with money and managing people. And then there's Jim Bell, a longtime friend. He is also the president of the board and a great source of insight, but the main thing we do for each other is make each other laugh. The three of them got me to take the job, and all three still offer great advice and insight and support. I cannot oversell the value of a support network of friends who can peer-review your own strengths and weaknesses.

I was with Neil, Dan, and Jim at a meeting in Britain, where The Planetary Society was presenting Stephen Hawking with the Cosmos Award for Outstanding Public Presentation of Science. Hawking showed up in person to accept the award and took the time to have dinner with us, no doubt because of the legacy of Carl Sagan. It was a great coup for The Planetary Society, but anybody there could tell that we were not taking the best advantage of the event. The British press should've been all over it. A few Planetary Society members who were invited were not dressed appropriately. Fine for them, I suppose, but it put me a little ill at ease. It was Stephen freakin' Hawking at the Oxford freakin' Library, for cryin' out loud! The youngest piece of furniture in the room was from the 1600s. The whole event just wasn't what it should've been; we weren't doing a good-enough job of telling an impressive, ambitious story about what we had all come together to do and to honor. My buddies on the board were thinking the same thing. The Stephen Hawking Cosmos Award evening could have been something much more memorable. This is when I realized the society needed a new direction or it might go out of business.

I had studied the society and concluded that it is the best nonprofit space-interest organization on Earth, but I also realized that I might bring the fresh perspective that the society needed at that point. So, with those three colleagues putting a little pressure on, I decided to take the job as CEO of The Planetary Society. I started going into the office before 8 a.m. every day because there were so many ways I wanted to make things better. Seven years later, the place is transformed. I'll take credit for it, with help, insight, and diligence from a lot of other people, especially my Chief Operating Officer (COO) Jennifer Vaughn. She and I have overseen a change of tone that, in my opinion, is seriously for the better. Some of the changes were painful. I had to take responsibility and lay people off. I had to hire new people with new skills. And I had to learn some new skills myself, including the arcane techniques of double-entry bookkeeping. Once I figured all that out, streamlined systems enabled me to control costs and ultimately get more things done. With the help of our longtime finance manager, Lu Coffing, I reworked the budgeting for our *LightSail*, the experimental spacecraft that rides the momentum of sunlight. The first *LightSail* flew in 2015, and we're planning to launch a second, more capable one to a higher-altitude orbit in late 2017.

Looking back, I can see that this process was really the flip side of the realization that everyone knows something that you don't: the realization that there actually were some things that I knew that other people didn't know, and there were some significant insights that I could contribute. As with every other kind of information, fear requires filtering. I needed to take a long hard look at my skills and home in on what

I was able to bring to the organization that nobody else was bringing. Then I could find the right kinds of experiences and inner strengths to call upon, and also figure out the best ways to talk about the things I wanted to do going forward so that people had a clear sense of where I was going. I didn't have traditional management skills, but in my engineering jobs, I had learned a lot about enlisting teamwork to complete complicated tasks. I had learned about budgetary challenges while working on *The Science Guy* show. I desperately wanted the society to succeed, and I had a clear vision for what that success could look like. Above all, I believe I had learned how to listen to others and how to recognize good information. I'm bragging a bit. But mainly I'm saying that we all have unique talents that we can bring to bear. Sadly, we can often be our own worst enemies when it comes to using the full breadth of our skills. Honestly identifying what your unique talents are gives you powerful protection against imposter syndrome.

If you are honest with yourself, it is also easier to judge whether other people are the real deal. Take the example of Elon Musk, the much-discussed founder of Tesla Motors and SpaceX. A lot of people love him, but he also draws plenty of skepticism. In the fall of 2016, I was at the International Astronautical Congress meeting in Guadalajara, Mexico, when Musk unveiled really far-out plans for sending giant rockets to Mars and building colonies there. He showed up with his amazing, photo-realistic electronic slides. I had to ask myself, as a guy who's been around these rocket and space people a lot over the past 6 years, do I believe he's the real deal?

Musk certainly does not seem to suffer from imposter syndrome, that's for sure. You can watch his startling large-scale presentation from Guadalajara on the SpaceX Web site. A lot of it strikes me as unrealistic. I don't think anyone will really want to colonize Mars any more than anyone wants to colonize Antarctica. A science base on

Mars is one thing, with geologists and exobiologists coming and going every few months. Otherworldly homesteaders living on Mars full-time, raising families for generations, with obstetricians and swing sets, is quite another. Then again, Musk may be way ahead of me here. It could be that I am not dreaming big enough. Certainly his ideas are inspiring, especially for the people who work for SpaceX and other visionary space contractors. Musk's employees come in at the crack of dawn and stay there until late at night. They may be the exact type of starry-eyed people needed to accomplish such a mighty thing.

Now, the saying, "Shoot for the moon; if you miss, you'll probably hit a star," doesn't entirely make sense when what you're discussing is actually traveling in outer space. But the gist of that saying is that you never know what you'll achieve if you dream big, and that's certainly the case here. As they work toward these goals, the people at SpaceX are already doing some very ambitious things. The company has what seems to be a great idea to reuse the bottom half of its Falcon 9 rocket. After launch, the first stage returns to Earth and fires its engines again to come to a soft, upright landing. SpaceX had a few rocky test flights, but now they can land those stages with great precision (most of the time). The plan is to use the first stages again and again, which should make rocket launches a lot cheaper. The engineers at SpaceX figure a booster can be reused about a dozen times, but nobody is sure if that is feasible. Will the company offer a discount to any satellite customer who puts their spacecraft on a rocket with the lower stage on its 10th or 11th reuse? Right now, the number of plausible reuses is low, and the cost of any failure is high, which limits the cost savings. This balance may change with more development and more experience. For example, if the bottom part springs a leak after a few flights, that is catastrophic. If it never does, that's great. When a rocket fails, the failure is usually explosive and complete.

Except for the space shuttle, whose costs were out of this world, nobody has ever tried flying the same rocket over and over. SpaceX will, and they'll get better and better at it. Does that mean SpaceX knows how to send people to Mars? Not yet. What I think Musk may not be fully addressing is how hostile Mars is. In addition, I question Musk's claim that we need to be a two-planet species to guarantee our survival. If something catastrophic happens to Earth, will we really turn to Mars as our safe place to run off to and rebuild? I think it's a great deal easier and much more practical to protect and preserve the Earth.

Weighing all the big talk against the actual accomplishments, I return to my question: Is Musk an imposter? My imposter-o-meter says definitely not. He created SpaceX from the ground up and is now competing ably with Boeing and Lockheed Martin. He runs the first company to have landed a rocket safely after launch. He's not shaking anybody down for money to fund his Mars dreams. I'm a skeptic about whether Musk will be able to achieve all the grandiose things he wants to do, but he has shown that he can work his way through the design pyramid from start to finish and do things nobody has done before. The man is the real deal.

❯ ❯ ❯

Full-on imposter syndrome—the phenomenon first documented by clinical psychologists Pauline Clance and Suzanne Imes in 1978—can be a debilitating condition. I don't mean to treat it lightly. But as I learned from my experience at The Planetary Society, a milder form of imposter syndrome is not entirely a bad thing. If you're riddled with insecurity, you will have a very hard time leading anybody. And if you have too little self-doubt, you can slide into self-delusion and hubris. Getting the imposter balance right requires filtering fear so that you keep things in proportion. That is a lifelong learning process. It reminds me of advice that the television host Tom Bergeron (you probably know him from *America's Funniest Home Videos* and *Dancing with the Stars*) told me: "Turn your nervousness into excitement." In the theater, in front of a microphone, while running a small nonprofit, or anywhere in life, a certain amount of fear means that you are taking a risk, challenging yourself.

There is another wonderful admonishment for stage and screen performers: "If you stop being nervous, it's time to quit." Nerves mean you are embarking on something daring and important. When you feel fear, you will know that you are on the right track. Let the fear come on, then take the time to reassure yourself that you can do this thing, that you can handle it. If need be, write down some of your favorite accomplishments that preceded this time of self-doubt. These touchstones need not be famous acts. Maybe you had a great time in a high school play, and maybe your buddy Rusty wrote that you were masterful. And you trust his judgment. (That happened . . . to me.)

If you are really good at internalizing and overcoming fear, people may never know that you felt it in the first place. Consider James Cameron. He went to the bottom of the ocean in his own $23 million submersible called the *Deepsea Challenger*. He wanted to do good science, but he wanted to prove himself, too. And of course, he wanted to go just because he's an explorer and the ocean had been a longtime fascination for him. Along the way he had some frightening moments: I'm sure he was filled with a bit of doubt when his ship drifted loose of its ship-to-ship cables and when he heard a very loud *bang* as the crushing pressure outside his vessel suddenly hammered his main hatch into place. A great many people were skeptical about his project because they thought of him only as a movie director. After *Titanic* came out, he famously proclaimed himself "king of the world," but that was the film world. In the world of underwater exploration, he was a relative unknown.

So Cameron decided to prove himself all over again. He designed his ship from scratch, and he was alone down there. He landed in the Challenger Deep, a spot 11 kilometers (7 miles) below sea level, much more carefully than the Navy did in an exploratory mission in 1960; he didn't kick up big clouds of silt everywhere. One fascinating thing he discovered is that a few kilometers away, where the ocean is only 30 meters (100 feet) less deep, there is an abundance of life. Where the *Deepsea Challenger* settled, though, there is virtually nothing living. Somehow, as the water flows to the bottommost part of the ultra-deep ocean, it gets starved of everything life needs: oxygen, nutrients, and minerals of the proper sort. We never would have known this remarkable fact if James Cameron hadn't decided he was an explorer as well as a moviemaker. He did some great science down there and brought a new perspective and profile to the work.

Even if you never try to travel down to the bottom of the ocean or build a rocket ship to Mars, there are two kinds of imposter syndromes that almost every one of us deals with and that are absolutely essential to overcome. First, there is the question "Am I just pretending to be a good person?" I hear this all the time, and I wonder about it myself. People wonder whether they are trying to do good for ultimately selfish reasons. Do I favor vaccines for the greater good or just to keep my kid safe? Do I support green energy only because it's the easiest way not to feel guilty about my comfortable lifestyle? Do I donate to a clean-water fund in the developing world mainly for the tax write-off?

When you take an everything-all-at-once approach, you can make those kinds of imposter feelings fade away. You slowly come to recognize that the things that benefit you also benefit the people around you. Reducing infectious disease, slowing climate change, and building up developing-world infrastructure all contribute to the global commons. In the long run, being selfish and being selfless take you to similar places. Nobody wants poverty. When people are poor, they are much more likely to cause crime; they contribute less to the overall economy; they suffer more disease. The extreme poverty in Chad has a strong influence on health. The average life expectancy there is just 50. You want everybody to have a decent standard of living. Even if you're a selfish bastard, you want everybody to have a high quality of life for the betterment of you. The best solution may not be the easiest, but it is ultimately in everybody's interest. That's why I talk about changing the world *for us*. In good design, everybody comes out ahead.

Then there is the second form of imposter syndrome, perhaps the most common of all. It's that feeling that the problems in the world—poverty, disease, climate change—are so huge, and the solutions are so complicated and daunting, that it feels absurd to presume that we can really solve them. It is easy to feel presumptuous even in trying; it's easy to give up and think, I'm not a truly good person after all because I'm not facing up to the hard reality. It's the flip side of being an actual imposter. When those feelings strike, remember one of the great gifts of nerd honesty. Looking at everything all at once doesn't mean that you have to solve every problem all at once. Let's face it—that's not going to happen. What you can do, though, is apply a rigorous standard to your actions. You can overcome the fear of being an imposter or in over your head. Do your ideas further your own interests by furthering the common good? A positive answer can emerge from small acts as much as from big ones. Your answers don't need to live up to some absurdly overwhelming standard—just a carefully considered one.

As you know, I love learning new words. Here's an especially relevant one that Corey Powell just introduced to me: "neltiliztli." It is an

Aztec term meaning "well-rooted, authentic, and true." It was their guideline for how to live a good life in an uncertain world on a sometimes-dangerous Earth, not by seeking power or affirmation but by doing your best to be in balance with your surroundings. The Aztecs—a society that most of us in the West do not associate with science—came up with a beautifully succinct expression for nerd honesty. The solution to feeling like an imposter is to be authentic, and the path of authenticity is the path to a better world.

CHAPTER 22

How High Can You Go?

Some years ago, I worked as a consultant to General Motors. I had made a deal that I would get to drive the EV1, General Motors' first electric vehicle, in exchange for making a few public appearances and sharing my feedback with the company's low-emissions designers and public relations people. I wanted to do my part to help move the market to zero-emission vehicles, and I was genuinely curious about what I might hear behind the scenes. In one meeting, a manager started his presentation with the statement, "We want our light trucks to be 50 percent recyclable." Well, that got me going. My immediate reaction was "No, no, no! You want your trucks to be 100 percent recyclable. *That's* the goal."

For me, it was a crazy-making moment. I barely even remember the rest of that meeting, but that one sentence still sticks in my head. It was like the manager was saying, "I sure hope I get a C in this class."

Here he was, in a roomful of bright, hardworking people who had made their way through the thicket of filtering, timing, self-doubt, and so on, only to get completely tangled up in confusion over short-term realities versus long-term goals. If you want to change the world—and creating a practical modern electric car is, and certainly was, a step in that direction—then you need to keep those two things very clearly separate. You don't want to set a mediocre accomplishment as your ultimate destination. It's okay to talk about small steps, but they should all be taken to get to a big goal.

It was unnerving enough to see that kind of confusion at General Motors, but for me it's been even worse to find it happening at the good old National Aeronautics and Space Administration. Here the problem runs in the opposite direction: not failure of imagination in the long-term goals, but unrealistic expectations in the short-term realities. At The Planetary Society, we are on a great many space and aerospace mailing lists. Every few weeks, or even every few days, I get e-mails full of high expectations from NASA's public information office. The messages are designed to solicit ideas or submissions for some new agency-sponsored venture, and they typically go like this: "This solicitation seeks proposals to develop unique, disruptive, or transformational space technologies currently at low technology readiness levels that have the potential to lead to dramatic improvements at the system level. Specifically, the proposals must address one of the following . . ."

Unique! Disruptive! Transformational! That is *not*, absolutely not, how actual technological progress happens, and if anybody should know that it's the scientists and engineers at NASA. Real progress unfolds bit by bit, small improvement after small improvement, with a clear, ambitious end goal in mind. Expectations of immediate, radical transformation are an expressway to disappointment and dead ends. They don't facilitate the long-term goals; more often, they prevent them from ever happening. I'm pretty sure the real researchers and inventors at NASA are fully aware of this. What's happening is that there is a management structure desperately trying to will break-throughs into existence simply by crafting an impressive-sounding project with lots of the right buzzwords and breathless e-mail solicita-tions. This is the shortcut-seeking behavior of an agency that feels pressure to perform beyond its resources.

There's a pretty good analogy to baseball. Your coach says, "Here's a bat, now go up there and hit a home run." It's not so easy, or we'd all be doing it all the time. Even making contact with a pitched ball is hard enough. Driving it out of the stadium is very, very difficult. At NASA, this disruptive, gotta-hit-a-home-run thinking is probably a leftover from the *Apollo* era. At that time, the agency's budget peaked at a little more than 4 percent of the US federal budget. That is a huge amount of money for any program. NASA was flush with cash, a young and talented workforce, and a razor-sharp directive to put humans on the Moon. Progress still happened incrementally, but there were so many innovations coming through so quickly that it often looked like overnight success—especially in retrospect.

Today the NASA budget is around 0.4 percent—that fraction is less than $1/10$th of what NASA's budget once was compared with the federal budget—and its mission statement lacks the clarity it once had. It may be that the managers, especially older managers, retain elevated expectations consistent with the resources NASA had in its heyday. More likely, the problem is deeper and wider. I think the program managers generally understand that the expectations are unrealistic, but they feel like they have no choice but to play along. This situation is hardly unique to the space agency. If a recording artist has a hit, the record label wants to know when the next hit song will be ready. Companies and people get a reputation for innovation and then the pressure mounts to keep the innovations coming. There are investors or friends or (in the case of NASA) congressional committees to be satisfied. If you are in danger of losing some of your funding, the pressure ratchets up even higher. And if you fail to deliver, the pressure on your next big promise ratchets up higher still.

This is where things get tricky for all of us who want to change the world. If you bite off more than you can chew, you won't be able to swallow. On the other hand, if you make plans that are too small, like that fellow at GM, you won't ever accomplish much at all.

❯ ❯ ❯

In case you haven't noticed, I have a big agenda for my change-the-world plans. I want to remake our transportation network, solve climate change, drastically reduce poverty, and vastly improve human health. Having a robust space program might seem like a secondary project compared with those things. I would strongly argue otherwise—pretty much the opposite, actually. Space exploration sets stretch goals that help us achieve all the others. It also forces us to think very clearly about how to get to those lofty end points, and address the short-term/long-term confusion that stymies everything-all-at-once planning—when it gets to the action stage. And, I'll admit, my interest is also partly personal. I'm a space enthusiast. I believe in the transformative power of exploring the universe. That's how I ended up as the CEO of The Planetary Society, where we emulate a lot of what old NASA was about . . . and hope to save some of new NASA from itself.

I often say that, around the world, NASA is the greatest brand the United States has. The acronym "NASA" is practically a shorthand expression for brilliant engineering, for the joy of science literacy, for the possibility of shooting for the Moon—for real—and getting there by meticulously carrying out all the amazingly difficult steps along the way. The agency is full of great scientists and engineers; even the managers and bosses, much as I feel concern about many of them, are unique. No other organization on Earth can do what NASA does. Nevertheless, there is clearly home-run hopeful pressure at work. For several years now, the agency has promoted a human "Journey to Mars" predicated on rockets that don't exist and life-support technologies that have not been developed. Much more important, those things have not been funded for development. So far, the journey consists more of beautiful graphics than real hardware.

But think for a moment what it would mean to have NASA unleashed again, with short- and long-term plans working together in harmony. Think what a real Journey to Mars could do for humanity. It could answer the most stirring question in planetary science, maybe all of science: Are we alone out here in the cosmos, or is there life in the universe beyond Earth?

If we found evidence of life on Mars—or stranger still, something still alive there—it would transform the way we think about life here on Earth. It would stir and shake people's philosophical and religious beliefs. It would galvanize science education. It would almost certainly affect medicine, biotechnology, and ecology, too. It would advance all the ways we want to change the world. But I think the hugest, everything-est thing it would do is change us. It would tell us that we are not alone in the universe.

I claim such a discovery would be as profound and world-changing as the proof that the Earth is round rather than flat or that the stars are other suns. The discovery of life would change the way every one of us thinks about being a living thing in the cosmos. This is the long-term goal NASA should be working toward.

We have the means to do this, and we have no shortage of system architectures and plans. What we lack are *good* executable plans, the kind that fulfill the principles of nerd design and planning, executed across all timescales. We're off to a decent start at least. Our Mars robots—*Sojourner*, *Spirit*, *Opportunity*, and *Curiosity*—are amazing and have transformed the way we think about the planet. The upcoming *Mars 2020* rover will be the most capable one yet, with radar vision and the ability to scan for organic compounds. But these machines, even though they are built by our best robot builders, piloted by our most skilled robot operators, and guided by our top planetary scientists, are not enough.

What we really want to do is send people. They will be scientifically trained, to be sure, but what they'll really be is explorers. When we send people out into truly unknown territory, two things happen. First, they make discoveries. Second, they have an adventure unlike any before. When people are orbiting above Mars or walking upon it, all the people of Earth will share in those experiences. It's been estimated that what our best robots do in a week, a human explorer (properly dressed and outfitted) could do in 5 minutes. If there is life on Mars, or a good place to go looking for life, astronauts will find it in relatively no time.

So why haven't we started? It's that short-term/long-term problem again. Humankind has kinda sorta gotten started on looking for life on Mars a few times, but there was always too much long-term vision and not enough short-term support. The Bush administration—the first one, under President George H.W. Bush—developed a scheme called the Space Exploration Initiative that would have sent humans back to the Moon and eventually to Mars. It had an unclear timetable and an estimated cost of $500 billion (in 1989 dollars no less) over a couple decades. In 2004, the second President Bush proposed an updated version, the Vision for Space Exploration, which would have put humans on Mars in 30ish years. It spawned an actual program called Constellation, which was supposed to create a new generation of giant rockets. But neither of those plans ended up going anywhere. They were beginnings without an end, with no long-term strategy.

There was a long-held presumption that once a NASA project got started, Congress would find ways to fund it, cost be damned. In reality, Congress took one look and declared the first Bush plan dead on arrival. After President Obama took office in 2008, he canceled the Constellation project, and the second Mars plan was gone, too. Obama took a lot of heat from space enthusiasts for giving up on Constellation, but the truth is that it was never even officially funded. You can't start creating rockets without a plan to complete them. It's something Ned Nye drilled into me and my brother and sister as kids: It's good to have initiative, but you have to have "finish-itive."

I feel that part of my job at The Planetary Society is to help NASA untangle its short-term/long-term confusion and come up with a better-balanced, realistic plan to do great things. Starting a couple of years ago, we proposed a program that would allow NASA to both search for life on Mars within current funding levels and accomplish its key goals on a politically meaningful timescale—over the span of a couple of presidential elections, say, rather than 20 or 30 years. Now we have that plan.

We started with this idea: To keep costs down and political support up, we have to use what we already have, the existing space technology from NASA and the private rocket companies. NASA's Space Launch System rocket is well along in its development and testing. SpaceX has the Falcon Heavy on the way, and another company, Blue Origin, recently announced a competitor called New Glenn. Let's not start over with new launch vehicles (rockets) yet again. I took lessons from my GM EV1 experience: no mediocre goals. But I also took lessons from the failure of NASA's previous Mars plans: no absurd home-run wishes. No calls for immediate "disruptive!" breakthroughs.

First, our team at The Planetary Society came up with an architecture that requires no real increase in the NASA budget. The budget, like that of every other federal program, would only have to be periodically adjusted for inflation. My friends, this is a radical idea in the Mars exploration community. Second, we set a hard date for humans reaching Mars: 2033. Third, we would start with humans orbiting Mars rather than going straight to the surface. This is how we went to the Moon. Humans orbited in *Apollo 8* before landing with *Apollo 11*. The human Mars landing would not happen until 6 Earth-years later, in 2039. The Planetary Society published these ideas in a 2015 study called *Humans Orbiting Mars*, which incorporated input from more than 70 experts at the Jet Propulsion Lab and other leaders in spaceflight.

To handle the cost of the mission, we proposed that NASA, per contract, spin off the International Space Station (ISS) in 2024, which would save at least $3 billion a year. There are other agencies and organizations that have expressed interest in taking over the operation of the ISS. All that newly available money could be redirected to starting stepping-stone missions back to the Moon, if need be, and on through cislunar space. I love that word, "cislunar." It means all the space between Earth and the Moon (literally "on this side of" the Moon), including stable points in the lunar orbit that might be extremely valuable spots for astronomy research. We crunched the numbers and found that there is a mission architecture that would work. We could get humans in orbit around Mars in 2033 without increasing the NASA budget.

While we are thinking about redirecting NASA, I argue it would be a big improvement—though politically very difficult—if we were to convert the eight NASA centers into what are called federally funded research and development centers. That would make them quasi-independent, with the ability to fire people, which, I can tell you, can change things in a good way for everyone involved. That's how the Jet Propulsion Laboratory in Pasadena, operated in conjunction with Caltech, works; it's also how the Applied Physics Lab near Baltimore, operated by Johns Hopkins University, is set up. It's no coincidence that those two labs have been responsible for some of NASA's most celebrated recent missions, including the *Spirit*, *Opportunity*, and *Curiosity* rovers on Mars, *Cassini* probe at Saturn, *Dawn* at the asteroids Vesta and Ceres, *MESSENGER* at Mercury, *Juno* at Jupiter, and the *New Horizons* flyby of Pluto.

The Planetary Society cannot tell NASA or the president or Congress what to do, but we do hope that the *Humans Orbiting Mars* report will serve as a guideline. Space exploration is beautifully non-partisan; it has no agenda beyond discovery, and it excites everyone regardless of where they fall on the political spectrum. The biggest obstacle is the perception that it is a luxury with little practical pay-off, and I'm working on that. From time to time, several members of The Planetary Society staff and I go to Washington, DC. For a full day, we meet with members of Congress. We show them the value of space exploration for science, technology, and education. We explain our methodical plan. Above all, we expound upon the remarkable discoveries that await us on Mars and beyond. One evening, the Society hosted a gathering of members and supporters at the Mott House, one block from the US Capitol. Congressman Adam Schiff, whose district includes Caltech, joined us and addressed the group. The mood was supportive. We will see. We are doing everything we can to amplify our influence.

I also have to say that as much as I believe in the search for life on Mars, it is not the only place where we can embark on new boundary-pushing ventures like the ones that made NASA so inspirational in its early days. Earlier I mentioned *LightSail-2*, an experimental spacecraft pushed by the momentum of sunlight, that has been developed and built by The Planetary Society. I hope to fly it in late 2017 or early 2018. NASA has a couple of inspirational missions of its own in the works. I was recently at Cape Canaveral watching the launch of *OSIRIS-REx*. (It stands for Origins, Spectral Interpretation, Resource Identification, Security-Regolith Explorer . . . I mean, really?) In 2018 the probe will visit the asteroid Bennu and bring back samples. Bennu is made of some of the most primitive material in the solar system. It is also on an orbit that brings it very close to Earth at times. *OSIRIS-REx* will allow scientists to learn much more about the origins of our planet and the rest of the solar system. The mission will gather data and test technologies that will be essential if we ever have to defend our planet from an incoming asteroid. NASA is also in the planning stages for a mission to Europa, Jupiter's giant ice-covered moon. Under its frozen crust, Europa has an ocean twice the volume of all the water on Earth. It's another fascinating place to look for life. If all goes well, we could be doing just that in the mid-2020s. NASA can still innovate with the best of them when the bureaucracy gets out of the way and the visionaries are allowed to set, meet, and plan both short-term and long-term strategies.

All of these missions and mission concepts are important not just because they push the boundaries of engineering but because they push the bounds of imagination. They are the quintessential perspective-shifting nerd undertakings. I already mentioned the implications if we find life elsewhere in the universe. But what if we don't? What if we keep looking and come up with nothing? That would be extraordinary, too. Either way, the answer will tell us something profound about our place in the universe: Are we part of a cosmic web of life, or are we something singular and unique? If it really seems like we are alone, that makes taking care of our planet even more poignant and profound.

I still hear the question all the time: How can we spend money on space exploration when there are so many problems here on the ground? I get it, but here's the thing: The problems are all linked together. The engineers at GM and the ones at NASA are dealing with the same problems—not just the same conceptual issues of short-term versus long-term thinking, but a lot of the same technical problems of battery design, control systems, autonomous navigation, and so on. It's no coincidence that Elon Musk's two main companies are SpaceX and Tesla; he clearly sees the connection. If we can reach higher in the search for life on Mars, we can also reach higher in developing clean energy, fighting poverty, and reimagining travel here on Earth. Let's go!

CHAPTER 23

Tragedy of the Traffic Accident

Like pretty much everybody in the world who drives, I've been in a few crashes over the years. So far, none have been serious, but every time something happens, I cannot help reflecting on the crazy inefficiency of the whole business. I pull out, watch for traffic, and someone rear-ends me; a guy runs a red light and smacks the left front end of my Volkswagen Beetle; or a car bangs into a taxicab I'm riding in. I drive close, obviously too close, to the guy in front of me. I almost avoid hitting his car, but I get rear-ended and pushed into his bumper, hard. All of these things have happened, and all of them make me think about how strange it is that we exert so much engineering effort to improve a system that fundamentally is so inefficient: all those cars on all those roads.

From one point of view, the setup we have now is vastly better than the one we had a couple of centuries ago. After all, we do move a great many of us a great many places at virtually any time of day and in almost any weather. Cars are much faster and more reliable than horses, and they don't litter our roads with manure. A lot of thoughtful design and engineering work went into all these roads, vehicles, traffic signals, and so on. But from another point of view, the way we do things today is just crazy. Look how big a car is compared with its driver. Look how wide a road is compared with a sidewalk. And look at all those specially-trained drivers in charge of heavy equipment moving at high speeds. How much of today's transportation infrastructure would be the way it is if we were starting from scratch? What if we took an everything-all-at-once, big-picture approach and allowed nerds to redesign it from the bottom of the pyramid up, focusing on safety and efficiency—how much better could they do? I have a feeling that they could do a lot better. Soon I think they will.

Our current transportation system has what civil engineers call a high "level of service" (ability to go place-to-place and door-to-door) using cars and trucks on roads—but only as long as we do not introduce too many vehicles, and as long as nothing goes wrong. The problem with too many vehicles is obvious: The need for adequate spacing between cars, the delays created when streams of traffic cross during merging and exiting on the highway, and the undesirable faster-slower waves introduced by vehicles moving at nonmatching speeds all conspire to slow traffic way down. And if you really want to screw things up on a street or highway, have a crash. They happen all the time, generally because fallible people like you and me are involved. Okay, no one like *you* is ever involved; you are an above-average driver who does a fantastic job even though every day you are surrounded by stupid people driving those other cars. I know, I know, it's your burden to bear.

But the statistics are sobering. In 2015 alone, 38,300 people were killed on US roads. Another 4.4 million were injured. You probably know someone who was killed in a car crash. You almost certainly know someone injured in a crash; you may have been injured yourself. With those injuries come hospital bills, lawsuits, loss of productivity, and, above all, personal misery—all things better avoided. This is the unsurprising consequence of putting hundreds of millions of people at the helm of metal-and-glass rolling road missiles, with only limited restraints to keep them from doing something truly stupid (like driving drunk or tired, or simply not watching the road while adjusting the radio or fiddling with a phone). We all share in the costs and in the misery.

For all that, it's still remarkable that the current system works as well as it does. The main reason is an embrace of fundamental physics. That is a tribute to the nerd engineers who mostly do the very best they can, notwithstanding the corner-cutting management mistakes made in a few clunkers like the Ford Pinto and Chevrolet Vega I railed against earlier. People still crash as they always have, but nowadays they generally walk away from the wreck; the auto fatality rate has been trending generally downward for decades because a whole bunch of engineers have been dutifully working within the constraints of our system to make cars safer when they inevitably do crash. What makes so many collisions survivable is our understanding of energy and how we can redistribute it through the car. You may have heard the term "crumple zone." It refers to the part of your car that is designed to literally crumple in an accident. Just as it takes a pretty good oomph to crush a beer can, it takes a lot of energy to crumple a car's extremities. Modern cars and sport utility vehicles (SUVs, aka family trucks) have been designed with crumple zones that absorb energy from the impact that would otherwise be transmitted to the driver and passengers. To comprehend just how well that works, I recommend you take a look at the process in action.

If you haven't checked it out already, look on the Internet for the crash-test video of a 1959 Chevrolet Bel Air running into a 2009 Chevy Malibu. (You can easily find it on YouTube.) It shows an "offset head-on" crash; that's when the driver's side of one car smashes into the driver's side of the other car. Many of us yearn for the good ole days of what seemed like overbuilt, too-strong-to-fail cars. But when you watch the videos of the test, shot from several different angles and points of view, it's shocking how much better the modern Malibu does at protecting the driver and passenger. By comparison, the Bel Air looks almost as if it had been designed to kill people. There was a lot of steel in the old cars, but it was not arranged especially well for a crash. It sent the crash energy right on through to the inside of the car. It was brawn without brains, you might say. Same brand, same basic type of midsize sedan, just built 50 years apart with a whole lot more of an understanding of how injuries occur.

All that great-looking sheet metal also wasn't arranged very well for the number-one job of a car, which is, of course, carrying people. When you lift the hood of one of these old vehicles and look inside, you see a great deal of unused space. If you're an average-size adult, and you were inclined to jump over the front grille, you could easily stand between the engine and the inside of either front fender. Some of that length is due to the excesses of 1950s styling, but much of it ended up there honestly. The total length of the car was largely determined by what was required to keep the crankshaft of the engine in line with the body of the car so that the twisting shaft could carry the power of the engine all the way to the rear of the car. Early on, cars were all rear-wheel-drive because it was a little complicated to have the front wheels do two jobs: the steering and the propelling. Nowadays, most passenger cars have front-wheel drive because the arrangement provides better traction and more room for the passengers inside. Better "packaging," as we call it in the engineering biz.

A modern car (or truck or SUV) is the most complex device you own; it's like your furnace, refrigerator, cell phone, and living room furniture all bundled together and put on wheels. There are only a few configurations that make everything in a car or truck work correctly, but there are near-endless configurations that would not work. You could accurately say the same about any modern industrial product. In design, chaos and disorder are easy to come by, while effective harmony demands an enormous amount of effort. That's why you should respect the good designs around you. Somebody went to a lot of trouble to arrange all the bits so that the thing—be it a city, a car, or a cafeteria tray—comes together to fulfill its purpose effectively. A good design requires a combination of broad insight and tightly focused attention to detail.

The everything-all-at-once approach—paying attention to impact forces, the power delivery from the engine, the exact shape and placement of the seats, and a thousand other details—requires a more complicated solution than arranging the engine and the rear wheels in pretty much a straight line. That was then. But today the added complexity of modern vehicles pays off with a vehicle that works and performs better. That approach has improved the assembly process, as well. And there's no need to restrict this insight to automotive design. The more details engineers take into account, the better the final outcome. Too many people overlook the vast practical payoff from decades of nerdy improvements in science and technology: Almost every product you use is designed better today than it was in the past. This is true of refrigerators, washing machines, ski jackets, bicycles, and windmills. Put a half-century-old vacuum cleaner next to a modern unit, and think about which one you'd rather clean your house with. We've progressed a lot since that 1959 Bel Air, in almost every way.

As it happens, my parents' 1963 Chevy Bel Air (the white one with the red interior that so charmed my sister and me) was not that different from the one in the crash video. One time, my father slammed on the brakes and my face smashed into the steering wheel. This was back before safety belts were standard; my engineering-minded father had added them, but not in the middle of the front bench seat, where I was perched. This was also, I should note, before my father invented his THANKS sign to signal other drivers. The impact changed the shape of the cartilage in my nose, and I couldn't smell so well until I got my nose professionally straightened after another trauma years later while I was playing ultimate flying disc (Frisbee). So I can tell you firsthand, those old cars were pretty brutal. Without a half-century of nerd engineering, they still would be. New designs needed to be only good enough in terms of safety. The marketing teams in the pyramid of design did their jobs well, a little too well, and buyers accepted the old safety compromises . . . for a while.

What really pushed the automobile forward was regulation. Much as Rachel Carson helped galvanize the environmental movement that showed up at the first Earth Day, Ralph Nader and a few other activists helped focus a public demand for better, less crash-prone cars. Nader has since gone on to become a bit of an extremist. But when he criticized the Chevy Corvair, people listened and agreed that we could do a lot better with a few laws to protect us all from carelessly engineered products. People asked for it, legislators demanded it, and the engineers were set loose to develop their better solutions. That's how we got our modern cars with crumple zones in the side and back as well as the front. The nerds looked at all of the constraints and came up with a good solution: Let the car crumple in a crash. In a collision, a modern car is often sent to the scrapyard, but car crashes are a lot less deadly as a result. A car that looks like an accordion after a serious accident is a sign that some engineers you'll probably never meet did their jobs very well indeed. It's been a long road to get to our current level of design refinement because designing a car is complicated, but that's only part of the story. The auto industry has also been held back by the kind of halfway thinking I encountered in my meeting with the GM engineers. (We still don't have an electric small truck, by the way.) The point is that we change the world when we want to. When we don't press for change, the results can be deadly. And in spite of all the progress, we have a long way to go.

Even as I celebrate the fantastic (phantastic) physics involved in keeping us from killing ourselves with our modern cars and SUVs, I cannot help acknowledging that all the fixes I've described so far are wasteful. Deciding as a society that we need cars that can crumple effectively means accepting ahead of time that we're going to be crashing and crumpling a *lot* of cars. That's accepting car crashes as a permanent constraint of the current system and the existing solutions. Now we (nerds, regulators, customers, all of us) need to take a look at that constraint to figure out a better way—not just to mitigate the danger of the constraint but to eradicate it completely. A better way to manage our transportation system already exists, in my opinion, and we all seem ready to accept it.

I'm talking, if you haven't guessed, about self-driving cars. It's an everything-all-at-once process, where you have to take the really big view to see the full implications and possibilities. Safety is the obvious benefit. At long last, we are about to really get on with it—stop merely trying to minimize crashes and start doing a lot better at preventing them entirely. Right now, flying is around 200,000 times safer than driving. Try this little thought experiment. You drive from Portland, Oregon, to Orlando, Florida, and back, let's say 10 times. It will take you a couple of months. Meanwhile, I fly round-trip from Portland to Orlando 10 times. Who do you think is more likely to get killed or injured? It's you, the driver. The planes themselves and the air traffic control system are way safer, hands down. So it will be with the self-driving cars of the future. Car crashes will still happen, but they will be rare, and they will happen in vehicles that are built to protect us even better than they do today. Engineers will keep working to improve safety every place they can. I'll bet that future generations will listen to old folks' stories about car crashes with head-shaking, decades-removed disbelief.

Once we have ultrasafe personal vehicles that don't require personal control, we will be able to start exploring novel ways to get around. You might not be a big fan of taking a public bus right now; maybe the route and schedule aren't right or there aren't enough seats. But what if you could summon a personal vehicle that would instantly show up at your door and take you to your exact destination? Would you mind so much if you were sharing that ride with a few others—especially if you could relax, work on your laptop, or play games on your mobile phone as you go? In this way, ride-sharing could catch on in a way it never has before. Remote locations that are not convenient to any kind of public transit service could become accessible by a summoned autonomous car. Even along well-traveled paths like Washington, DC–New York–Boston, you might find it faster and more pleasant to relax in a quiet, electric self-driving vehicle than to haul yourself to the airport and hop aboard an airplane or make your way to the Amtrak station. It will certainly be better than piloting your own car along the crowded lanes of I-95, and more efficient, too.

From there, I have no trouble imagining a world in which only a small number of people even bother to own cars. Automobile manufacturing will change completely. Instead of stylish new models arriving every year, long-lived robust body shapes, and spare parts for them, could become the norm. Most of the time, transportation will be a service that you activate as you need it, like cable TV or Internet access. You might still drive for pleasure now and again, but mostly you'll be free to let transportation companies and municipalities deal with all the inconveniences of owning and maintaining cars. Through my engineer's eyes, it makes a lot of sense. Changes in behavior will allow changes in infrastructure. For instance, we could rebuild the streets in dense urban centers so that narrower self-driving taxis ("taxipods"?) of the future can roll closer together than our extra-wide vehicles of today. Parking spaces could become new pedestrian or social areas. We can reclaim the domestic real estate currently devoted to garages and driveways. Many of today's old assumptions may become part of tomorrow's charming old anecdotes.

Some people wonder if the transportation world is ready for this kind of a nerd makeover. Personal car ownership is so deeply ingrained in Western culture—American culture especially—that changing it seems radical, maybe even unnerving. Then again, I work right now with several young people who don't own cars. They take mobile phone–summoned livery service cars everywhere they go. To me, the current situation with automobiles is analogous to what happened with horseback riding a century ago. A small number of people love horses and still spend some of their time and income on riding and taking care of horses. But for most of us, horses are just not as comfortable or easy or capable as cars and subway trains.

Think of the productivity we will gain when we're no longer spending hours per week manually navigating from place to place. Think also of the productivity we won't lose. We will steer clear of untold pain and suffering, hospital visits, insurance costs, legal fees. I won't have to drive around worrying about getting rear-ended, or smacking someone's bumper, or watching my car get crumpled by a texting teenager. When we don't have to pay attention behind the wheel, we will gain 45 minutes a day, on average, for every American adult. Lost time will become found time when we can engage our big brains in reading, writing, creating art, playing games, connecting with friends, doing all kinds of jobs that don't require manual labor, and enjoying all kinds of modern forms of recreation. We'll have more time for contemplation of everything all at once and the positive-feedback loop that comes with it: With contemplation comes the nerdy inspiration that will lead to the next automated systems to make our lives a little better still.

Now if you don't like the idea of automated travel, I hope you're sitting down, because I have some bad news: When you boarded the airplane, you trusted an automated system there, as well. Most of the time, the "pilot" in control of a modern airliner is the autopilot. Or autopilots, plural: There may be three autopilot systems that vote, in the electronic sense, and if they don't agree, they beep (loudly) and the human pilot has to take over. Autopilots can keep a plane flying straight and level, from point to point, with no human involvement; they can carry out the approach and landing; and they can deal with emergencies through every phase of flight. These systems are based on even more capable versions developed for military planes. It reminds me of a joke the engineers used to tell at Boeing. "Have you heard about the B-3 bomber? It has a pilot and a pit bull. The pilot is there to watch the instruments. The dog is there to make sure the pilot doesn't touch anything." So far, there is no B-3 bomber with these features, but this joke is funny (to me) because it's not that hard to imagine. Incidentally, I feel the same way about many of the leaders and officials with decision-making capacities on the big issues of our day. They may be the pilots, but we have to be the pit bulls.

I'm all for making our systems as automated as they can be, so long as we are the automators. If we let our machines become unsafe, it is not the fault of the machines—that one is on us. If we let our governments run amok, it is our fault there, too.

Judging from the public fascination with early self-driving-car experiments and their rapidly proliferating offspring, it won't be long before we are all willing to let go of control on another level, far beyond trams and airplanes. Soon we'll make peace with the idea of being hands-off passengers in our own automobiles, even when we are driving alone. Cars will be like airplanes in one way, but quite different in another. Right now, air traffic controllers (the folks working in those towers at the airports), and the systems they operate, direct and control airplane flights. They make sure the planes stay out of one another's way; they set each plane in place for takeoff, and they line the planes up for landings while they're in the sky. The air traffic control system is largely top-down: Air traffic is monitored and directed by centralized computer systems, and the planes are dispatched or held back according to a master plan.

Self-driving cars and trucks will almost certainly not work like that. Instead, people like you will get in their cars and command the cars where to take them. For the most part, we're still very much in the design phase of imagining the world of self-driving cars, but it's getting closer. We are improving our systems, nudging closer with every passing mile. As I write, Uber is already rolling out an experimental fleet of such self-driving cars in Pittsburgh. Many other similar taxi networks are coming soon, and those will be followed by self-driving vehicles sold to individual owners. There are still a lot of problems that need to be solved before we're ready for that, but engineers around the world are hard at work identifying the constraints and finding ways to overcome them. Companies are working their way up the upside-down pyramid of design. In short, this will need an everything-all-at-once perspective. We will need to evaluate the designs critically at each stage and change our minds when we find that some approaches don't live up to the hype. And generally, we will have to apply the rigorous standard that an autonomous car will have to do it better than you or I can . . . all the time, every time it takes you into traffic.

If you look back, you'll realize these ideas aren't exactly new. At the 1939 New York World's Fair—a time when my parents, Ned and Jacquie, were a young college couple in love, and my existence was not even being considered—the Futurama exhibit featured autonomous, radio-guided electric cars. The only problem with the cars was that they weren't real. They were an inspired vision from an industrial designer named Norman Bel Geddes, the mastermind of Futurama. The technology of the time was nowhere near what was needed to make such cars a practical reality; it took reality more than 70 years to catch up with the optimal solution (or at least a more-optimal solution) that nerds like Bel Geddes could see so tantalizingly in their imaginations. But you know what's cool? Even though it took a few generations, we *are* catching up.

Whenever I look down from an airplane window or look out from an upper floor of a tall building, I am struck by the colossal waste that surrounds us. We're using human brains, precious biological systems capable of closing keys on a clarinet or computing the curves of calculus, to accelerate and brake and steer our cars . . . on unguided roads, where a single twitch or a moment of inattention can be fatal. As an engineer, I've longed for something better than take-them-as-they-come solutions like bumpers and crumple zones. Now it's really happening. Manual cars will vanish from day-to-day life because they are part of a system that we will soon recognize as almost comically dangerous and inefficient. We will embrace better design, as we have over and over again in the past.

The first self-driving cars will not be entirely autonomous, and their systems will be only as good as the humans who designed them, but, oh, what a start. They will improve, fast. It's already happening. Then, after a few engineers crunch the problems tied to self-driving—self-navigation, self-collision-avoidance, and self-who-goes-first-at-the-intersection-agreement—the world will be easier on us. These cars will take a lot of the aggravation out of our driving experiences in stop-and-go traffic or long, monotonous stretches of highway. They will greatly increase the mobility of people who are physically disabled or elderly. They will provide safe rides home to people who had too much to drink at dinner. Autonomous vehicles may still have a setting that lets you grab the wheel, but they probably won't let you drive like a crazy person darting in and out of lanes to get where you're going. There won't be a need for RoboCops of the future; the system will have other vehicles box you in if you start driving wildly.

Getting used to the idea of self-driving cars will require a big shift in thinking for all of us drivers who are used to being in control of our cars, but we'll be giving up that individual control for the greater good. And it's a completely precedented move: We do it on planes, on subways, and every time we get in a car with another driver right now. It's not a burden; it'll be just how we roll. Put another way: Our technology is going to enforce a dose of nerd politeness and responsibility. It is going to make us all take better care of one another.

CHAPTER 24
Cold, Hard Facts of Ice

In the summer of 2016, I visited the East Greenland Ice-core Project (EGRIP), where a group of climate scientists are drilling far down into the Greenland ice sheet—a 400-kilometer (250-mile) wide book with pages made of snow.

My visit to Greenland was an emotional experience. Over the past 2 decades, I've reviewed the irrefutable evidence that humans are influencing the climate around the world. I've spoken out about the issue loudly on television, in public lectures, and in books. Still, like most people, I've found it hard to think about what is happening in an immediate, visceral way. The world is too big for most of us to think about every distinct ecosystem, and climate change is a complex, subtle process. Then there are the scales of time: hundreds of thousands of years. I mean, what was anyone, if there were any ones, doing a thousand centuries ago? It's wonderful and baffling to contemplate. Our lives are so short in comparison. When you visit Greenland, most of those abstractions go away. The variability and sensitivity of the global environment is laid bare. There is no denying the possibility of sudden and catastrophic change. There is no question about the urgency of action. We can look deep into the past to anticipate the future, using the predictive power of science. That's what the researchers are doing on the Greenland ice.

The evidence I saw is deeply troubling.

Greenland holds a comprehensive record of Earth's ancient climate, because everything that happens there gets naturally stored; it's literally a deep freeze, over a mile deep. The evidence here is so pure that scientists have to do very little filtering to grasp what's going on. Although I visited in summer, the temperatures atop the middle of the ice sheet never rose above -5°C (23°F). During what would be nighttime hours, it can get quite a bit colder, down to -20°C (-4°F). In this case, the word "night" refers only to the hours on a clock. For months around the summer solstice, the Sun never sets, and for months in the winter, it never rises. During my trip, I woke up around 3 a.m. and took some pictures because it's almost as bright outdoors then as it is at 10 in the morning.

Each year, new snowy layers accumulate on top of the glaciers that cover Greenland, forming enormous 100-kilometer-long sheets of solid ice. The snows become a record of the atmospheric conditions of the time. Variations in temperature and humidity produce flakes of different sizes and thicknesses. Depending on the winds, there might be a dusting of soil blown in from another continent mixed in or on top of some of the snowfalls. Over the course of a season, the accumulated snow forms a record of what the environment was like during that season. Winter snows and summer snows have slightly different qualities. In Greenland, it has been snowing every season for hundreds of thousands of years, and the snow does not melt away, so we are talking about a *lot* of snow—piled up and up. The snowflakes in each layer get squeezed together by the weight of the additional snow above. The air between the snowflakes has nowhere to go, so it ends up in tiny bubbles of atmosphere that are trapped between the tines of the flakes. The Greenland ice sheet is composed of these compressed flakes: more than 100,000 piled-up seasons of snow preserved by the cold. That's what ice researchers mean when they call it a book of snow. It is a natural ledger of the winter-summer cycle of seasons and of the ancient atmosphere and climatic conditions of Earth.

Over the last 40 years or so, engineers have developed specialized ice-coring tools to lift these pages of climate history to the surface so that we can read them. It is, as you'd imagine, hard work. The EGRIP work is sponsored by the University of Copenhagen, so the food there is prepared in the contemporary Danish style, which is to say, fantastic; that keeps everyone energized, and boy do they need to be. Everyone there—the drilling crew, paleoclimatologists, carpenter, electrician, doctor, and cook—is contributing to get the ice cores to the surface and to preserve them for study. The small drilling crew has an especially difficult job. To stay out of the wind and out of the sunshine, they carve out a hollow workspace for themselves using ski resort–style tracked vehicles. They dig trenches 30 meters (100 feet) long, 10 meters (30 feet) wide, and 10 meters deep. They inflate enormous hot dog–shaped balloons in the trenches. Then, using snow blowers, they create a roof by covering the top of the balloons with about 3 meters (10 feet) of the surrounding snow.

The result of all that work is an enormous ice cave. The finished under-snow hollow space looks like the lair of a comic-book villain. Down in the ice tunnels, you are protected from the elements, but it's hardly cozy. I'm not kidding when I say it's freezing. It's actually colder down there than on the surface. Everyone has to move around cocooned in multiple layers: warm long underwear, thick insulated pants, massive insulated boots, and usually a very thick down jacket. Up top, you need to wear sunglasses to deal with the blinding light reflected off the ice. Studying climate change is no leisurely pastime. You have to really want the cold, hard facts to do this kind of research.

Out there on the vast frozen expanse, the EGRIP team used surface-penetrating radar to find where the ice sheet is thickest. It's where they expected to find the most layers of ancient snow created by the most years of snowfall, and therefore the longest, oldest record of ancient climate. To get the best data, the researchers selected this particular location in the center of the East Greenland ice sheet. To get to the good stuff—the really old, deep stuff—they lower a remarkable custom-built drill-tube into the extremely solid ice. The tube is fitted with an electric motor, which drives sharp cutting bits. The crew lets the machine bore down, creating a long, thick cylinder of ice; the drill tube also has spring-loaded "dogs," metallic paws that grab the end of a captured ice cylinder so that it can be winched up to the surface with its steel cable tether. In the cold work area around the drill, research-ers carefully saw the extracted ice cores into standard 55-centimeter (22-inch) lengths, weigh each one, and note any obvious imperfections that might have been created by their coring system.

All the drilling, sawing, weighing, and measuring make up just the start of the process. Scientists examine the chemistry of the ice and the entrained air in great detail. We can know for certain how many years ago the snow fell and what the climate was like at the time. We can know what the atmosphere was made of back then and where the dust embedded in the ice came from. We can read the pages of the climate scientists' history book. By tracing your finger along an EGRIP core, you can trace your way through a passage of ancient time. I did; it's amazing. Perhaps because I can talk the engineering talk (and maybe just because I was there with a film crew—some disclosure), the EGRIP team let me handle several of those precious cores. These things are really hard to replace; they're priceless, in their way. Each comes up as a cylinder of ice about 2 meters (6 feet) long. That much ice is quite heavy, but when I was down there handling these things, it was so exciting that I didn't notice.

Whether in the very cold hollowed-out cave laboratory connected to the drilling area or back at a similarly cold lab in Copenhagen or Denver, climate scientists count the layers of snow ice. They do it just like you would count tree rings in a tree stump, and it tells the same basic story: Each layer or closely packed group of layers corresponds to 1 year (one season) of snowfall. If the separating layer between years is not optically obvious, the researchers examine the ice more thoroughly using a pair of electrical conductivity probes, which can detect subtle year-to-year differences in the composition of the snowfall. Not too far from the surface, where the old snow layers are less compressed, trapped air bubbles are readily visible. Deeper down, the layers corresponding to each snow season are squeezed thinner and thinner by the increasing weight of the snow above. The bubbles are still there, but they get compressed nearly down to invisibility. Deeper still, the bubbles are squeezed to such a degree that they disappear; they've dissolved completely into the solid ice.

I was there long enough to watch the drilling crew get past the relatively low density "firn," firmly packed yet obviously still-porous snow. I was helping (or trying to help) as a core from the year 1889 came up. All the researchers immediately recognized that layer as the core was laid out on the measuring bench; that is how well they have come to know the ice. I could see a distinctive, thin hard line indicating that it had been a particularly warm year. As I write, using the ice and a great amount of data gathered from around the world, 2016 was shown to be the warmest year on record. It's part of the 250-year postindustrial trend. The ice carries proof of humankind's global reach.

After the first day of drilling of the season, we all shared shots from a bottle of excellent Scottish whiskey, celebrating a new line (or cylinder) of research and a challenging job begun. When pieces of ice from great depths are brought up to the drilling workspace, the embedded air is ready to burst out with the force of the released pressure. It's fun to grab some of the waste ice chips that are held in the outer barrel of the drill and put them in your glass of Scotch or whiskey. The ice sizzles as compressed bubbles from Earth's long-ago atmosphere reemerge after thousands of years.

The most important thing to know about EGRIP research is that it does not merely record the past. It also provides ground truth (ice truth?) about what to expect in the future. It pits all the ideas and spurious theories purported to show that humans aren't affecting the climate against cold, hard, absolutely irrefutable evidence. Along with the scary proof of modern human-caused climate change is another very frightening phenomenon. The Greenland ice contains detailed evidence of what ice researchers are calling "abrupt climate change." Now, people, this is serious. Over very short periods, snowfall and rainfall patterns changed. Storm patterns changed. Ocean currents changed. We're not talking about "short" on a geologic timescale. We're talking about substantial climate shifts unfolding over decades, or even just a few years. If one of these abrupt climate-change events happened when you were born, by the time you graduated from high school, the land where your food had been grown might be completely barren. Can agricultural systems be moved fast enough to feed everyone? It's a disturbing scenario described to me by climatologist Jim White of the University of Colorado. These abrupt events can happen embedded in what we perceive as the slow overall pace of natural processes. Today we are poking at the Earth's climate system much faster than any natural phenomenon you can think of. At some point we may well run into one of those abrupt transitions in the climate system, and we don't know when. What the computer models cannot (yet) predict is in some ways even more frightening than what they can.

Ice ages are dramatic examples of just how extreme climate change can be. We are living more than 10,000 years since the last great cold spell, but evidence of past ice ages is still all over the place. I went to college in Ithaca, New York, which is on Cayuga Lake. My first job out of engineering school was with Boeing in Seattle, which is on Lake Washington. My high school pal Brian lives in Cleveland, which is on the shore of Lake Erie. All these bodies of water were carved by glaciers. Cayuga Lake, like the rest of the Finger Lakes, runs north and south, as does Lake Washington. They all track the flow of long-gone sheets of ice scraping down from the cold, cold north. As those sheets moved, they carved out lowlands that became the lakes. The Great Lakes were formed from "dead ice"—enormous chunks of glacier that had stopped flowing downhill, their massive weight pressing down into valleys that had been gouged and enlarged by the once-moving ice.

The processes that triggered those dramatic events were remarkably subtle. The primary causes of ice ages are changes in the shape of Earth's orbit, and in the degree and direction of our planet's tilt relative to the Sun. These changes run in cycles of about 100,000 years, 41,000 years, and 23,000 years, respectively; they are called Milankovitch cycles, after Milutin Milankovitch, the Serbian mathematician who discovered them around 1912. Nowadays, though, the climate is changing much faster than any Milankovitch cycle. Our situation demands urgent, measured action.

When the effects of the cycles combine to put slightly less than average sunlight on the Earth's surface, our planet cools a little. When the cycles align to deliver more sunlight, the Earth warms a little. The chemistry and circulation of the oceans amplifies these effects. When the Earth warms, the oceans release some of their dissolved carbon dioxide and water evaporates more quickly from the ocean; added CO_2 and water vapor in the air exacerbate the warming. Conversely, when the Milankovitch cycles combine to deliver less sunlight, Earth cools a little, the ocean absorbs more carbon dioxide, and less water vapor makes its way into the air. That amplifies the cooling and you get an ice age.

To get a proper nerd perspective on climate change, what you would really want to do is run experiments on the planet to see how it behaves—to study cause and effect. We can't do that, so we build computer models and compare their output with the data in the ice. We see if we can write software that predicts the past and models the facts in the ice. We can check our assumptions and biases.

Meanwhile, when we look through the ice cores from Greenland and elsewhere, we see abrupt changes in the layers. That is the proof that major changes can happen in less than 20 years. We still don't know exactly how those changes happen, but we have a damn good idea why: It's the interplay of temperature, carbon dioxide, ocean currents, water vapor, and the living things that respond to it all. There are a few other important factors, as well. If there is less winter snow and ice, Earth's surface is darker overall and absorbs more sunshine, which leads to more warming. If more water melts off the Greenland glaciers, that changes the balance between freshwater and saltwater in the Atlantic Ocean, which can alter its circulation pattern. Ocean currents distribute heat around the planet, so any shift there can have great big consequences. These are the processes that humans are tinkering with by burning fossil fuels, clearing forests, and otherwise amplifying Earth's greenhouse effect. The year 2016 was a milestone: The atmospheric concentration of carbon dioxide worldwide went above 400 parts per million *for the first time in 4 million years*. People caused it, and people will pay the price. The main questions are which people, and just how high will that price be?

We are in uncharted territory, which is why it is urgent to start moving toward a carbon-free future—and to start doing it now. The men and women of ice-core research describe Earth as a chaotic system. Roughly put, a chaotic system has inputs and interactions that can lead to unpredictable and sometimes enormous changes as a result of tiny disruptions. You may have heard of the "butterfly effect," the idea that the flutter of a single butterfly in South America can eventually cause a hurricane to start spinning off the west coast of Africa. Well, we may be facing that kind of effect for real. That is part of the story recorded in the Greenland ice. Humans are taking on the role of the butterfly—but instead of a tiny wing flutter, we are dumping carbon dioxide and methane (an even more powerful greenhouse gas) into the air a million times faster than nature does.

Sometimes the self-proclaimed skeptics point out that climate has always changed. They are right—but only up to a point. It has always changed, but the *rate* of change that humans are introducing now is unprecedented. You can think of it this way. When you are driving on the highway, you might be going 70 miles per hour. But when you arrive at home, your speed is 0 miles per hour. And look, you are doing just fine. But note well: You could also go from 70 to 0 by crashing into a brick wall at highway speed. That would lead to a bit of a different outcome. Imagine hurtling down the highway with someone else at the wheel. You see a brick wall built right across the road up ahead. You yell, "Hey, slow down." Then the driver says, "Don't worry about the brick wall; car speeds are changing all the time." It ain't the change that's the problem, my esteemed colleagues—it's the rate of change that gets you. It's very much the same thing with the climate. And imagine, even as the wall approaches, that same someone further telling you, "Don't listen to those car-crash experts, they are just trying to get rich with those physics grants."

These days, it seems as though for every genuine nerd expert, there is someone else peddling false doubt. That's why we have to work together, do some thoughtful information-filtering, and make the informed voices heard. We need to make sure everyone understands the real situation; the welfare of the planet (and those of us who happen to live on it) hangs in the balance.

While I was up there on the Greenland ice, I got the creeps. And I don't mean the creeping movements of the ice sheet, which slides about 15 centimeters (6 inches) a day. I mean, my stomach did a flip when I thought about the consequences of the discoveries being made there. As goes Greenland, so goes the rest of the planet. The rise and fall of ice ages recorded in those half-meter (22-inch) EGRIP ice core sections reflect ancient climate changes that were, and may be soon again, buffeting the entire world. The rapid warming now under way will be felt around the world, too. Despite occasional heavy snowfalls, we will probably see more droughts in California, summer heat waves across Europe, and catastrophic flooding in South Asia. There will surely be other butterfly-effect consequences we haven't even considered yet. This is the tragedy of the commons in the extreme. No single person thought he or she would or could have an effect on the entire planet.

But all of us carrying on as we are, with business as usual, will all soon feel the effects in big ways.

Researchers predict and estimate the possible consequences. Drastically altered rainfall patterns may lead to repeated failures of crucial crops—rice, wheat, corn, soybeans, and cotton. Warmer winters could allow insects to spread. They may destroy crops or spread heretofore-tropical diseases like malaria and dengue fever to London, Moscow, Tokyo, and Minneapolis. Major urban areas may run into shortages of drinking water. And without adequate ways to get rid of their sewage, all sorts of diseases may emerge. In the extreme case, climate chaos may unfold in such an unexpected, rapid, catastrophic fashion that it becomes difficult to prepare suitable large-scale responses. Even if the warming unfolds in a more predictable way, such that we can more easily foresee the food shortages, electricity shortages, fires, droughts, heat waves, and sewer system failures, we still probably won't be able to stop all or even most of them from happening.

Climate change is the great test of our ability to harness critical thinking. So far, we are pretty much flunking this course. We need to do better. We need to pull out the best of everything—data, design, and execution, which will be shaped by our sense of collective responsibility—and get to work. We need to manage our greenhouse gas emissions. We need to provide clean water and reliable, renewably produced electricity for everyone on Earth. We need to hustle because the longer we wait, the worse the problem gets. The time to start is now. This problem is bigger than any one country and any one administration. If our leaders don't take collective action soon, then we have to become leaders ourselves. If deniers and obstructionists make confusing noises of misinformation, we have to make our nerd voices heard more loudly.

When future generations look back at the record of Greenland ice from the 21st century, I want them to see the year when we sprang into action. A course correction of that magnitude will require a broad, sustained worldwide effort. To save the planet for us humans, we will have to pay attention to our shared interests rather than stumble into chaos as unconnected, self-interested individuals. We have to harness both knowledge and responsibility.

We have to change the world on behalf of all the species—especially our own.

CHAPTER 25

West Virginians and All That Coal

In October 2015, I was invited by the West Virginia Higher Education Policy Commission to present a lecture about science and the environment in Charleston, West Virginia. A central point of my presentation was the need to embrace clean, renewable energy. This part of my presentation always provokes strong reactions. Sometimes people laugh, and sometimes they nod their heads as I make a serious point. Better yet, perhaps, sometimes they shake their heads in violent disagreement, which means I'm getting a response and maybe even evoking reconsideration. I was prepared for plenty of the hostile variety of head-shaking when I spoke to the West Virginia crowd. This is coal country, after all, and I have some highly critical things to say about the black rock.

For those of you who haven't visited Charleston, it is a beautiful place, much in keeping with the state song, John Denver's "Take Me

Home, Country Roads." Then you drive down a genuine, postcard-perfect example of a country road and immediately lay eyes on—an aggressive coal-mining operation. Gazing down from the airplane on my flight in, I saw enormous open gray rocky areas, like barren islands in a sea of green trees. In some places the gray completely dominated, as if someone had dumped an enormous bucket of industrial paint over the landscape. All around there were muddy lakes, patches so drab that I wasn't sure they held any water at all until I caught sunlight glinting off them. And those lakes aren't clean water; they are tailing ponds—giant outdoor slop buckets of mining fluids.

To anybody with even a hint of environmental sensitivity, the scene would evoke despair. How could we have destroyed so much forest? How could we dig up ancient toxins and leave them sitting there on the ground? How could we have wrought such devastation on those gorgeous forests and Appalachian peaks in the span of less than 2 centuries? We built roads, ships, factories, and cities with the fossil fuel energy, but many of the people who once loved the land can no longer live upon it. No farming or hunting is possible near the human-made lakes of industrial goop, and drinking water is contaminated in many areas. Those long-ago entrepreneurs didn't understand or didn't care about the short- and medium-term, let alone long-term, consequences of their actions. Companies came in for enormous profits. Politicians made decisions that protected jobs for the next 2 or 4 years—not the next 20 or 40. Now the only coal that's left is not so high quality, nor is it so easy to remove. The ends that justified these means are fading away, as clear an example of short-term gain in exchange for long-term loss as you'll ever see.

I tried to maintain historical perspective as I contemplated that bleak scene. The whole-scale alteration of the natural landscape is a testament to technological innovation and the power of human industrial might. Earlier generations of nerds figured out how to harness the chemical energy stored in the once-living green plants of ancient swamplands—plants that got buried millions of years ago, compressed, transformed by heat, and converted into coal. Industrial-age innovators came up with ways to extract the coal, using it to power steam engines or converting its stored energy into the fossil fuel–powered electricity that drives so much of our modern world. Those were true nerd-genius moves. Many of those long-ago scientists and engineers were following the best design practices and working with the best-available information. They were trying to make a better world, too. Ultimately, theirs was a short-term solution. We have to change.

For about 2 centuries, the guiding approach in the industrialized world was essentially: dig, dig, dig. Burn, burn, burn! Earth's natural resources looked enormous compared to human demands. As many a stranded driver has experienced, though, you *can* run out of gas. It took nature millions of years to create the coal (along with oil and natural gas). Once those resources are gone, they're not coming back in anyone you know's lifetime. Sooner or later we'll have to turn to the new, better energy sources available to us—and the longer we wait, the worse things will be for everyone everywhere. We all share the air . . . But fossil-fuel energy technology is well established, and the profits from burning are solidly guaranteed. There are enormous political and social forces motivating industry to stick with bad energy policies that are based on outmoded perceptions of Earth's capacity to absorb carbon dioxide.

In this region, as technology advanced, and miners removed the most accessible coal, the mining process got more and more disruptive. Even if you aren't familiar with the word, I'm sure you can conjure up a mental picture of an adit, the entrance to a mine. ("Adit" is a traditional crossword puzzle and Latin-class vocabulary word: the opposite or complement of "exit." Add it to your φ phile.) An old-school coal mine's adit leads straight into the hill or mountain that contains a coal deposit. The coal layer is often nearly horizontal, a coal "seam," like the bottom of the calm pond that may have once rippled where the coal was formed millions of years in the past.

You can still see adits and mining tunnels on a tour through West Virginia, but they are relics. Nowadays, supersized mining equipment, shovels, and trucks have taken over. That is one big reason for the 10,000 lost jobs in the last 5 years: mechanization. The modern machines are huge. How huge are they? A dump truck can be three stories tall. Most of the time, it is easier to use these machines to remove the whole mountain than to tunnel into it. Mining companies coined the unsettling term "mountaintop removal" for this new style of getting at the coal. You do not have to be a miner to realize the first big problem: What do you do with the mountaintop itself? What happens to all the wildlife, forest plants, birds, bees, and trees that used to live there before you arrived? What happens to them after you leave?

The answers ain't pretty. Everywhere the mines are dug, the "overburden"—the entire mountain that used to sit above the coal seam—is crushed and generally dumped into the valleys below. That dumping creates some of the treeless areas you can see from the air. Those areas were once forests with canopies, understories, and floors that provided homes to swarming insects and crafty trout. At the fertile bottom of these valleys is where the streams flow to nourish the wild ecosystem and provide drinking and tooth-brushing water for the humans living nearby. Or rather, it's where they used to flow. Minerals from the mountain rubble leach into the streams, often rendering them toxic. Local fish and bird populations have taken big hits. People living nearby suffer elevated levels of cancer and heart disease, possibly from toxins in the water and air. And it costs a lot of money to relocate residents whose homes and property have been contaminated or outright destroyed. It ain't cheap to ruin everything. Mountaintop-removal mining externalizes costs to all of us.

I can see how someone living in West Virginia could feel deeply conflicted by all this. The coal-mining business here runs back to the middle of the 18th century. For many families, it is a tradition that has spanned multiple generations. Even in its reduced state, West Virginia still has 51 minable coal seams producing 60 million tons of coal from surface mining and another 80 million from underground mining every year. More than one-third of the electricity in the United States comes from coal; globally, the number is even higher, at over 40 percent. Coal is important to the modern world. Problem is, it is too important. Coal is the most carbon-intensive fossil fuel, and coal burning is the number one source of human-generated greenhouse emissions. These are not matters of opinion or politics; these are facts.

Not to mince words: The future of the Earth depends on coal not having a future in our energy supply. If we want to avoid spoiling our global commons, we need to move on and find better ways to get our energy. That is the tough message I was bringing with me to West Virginia, and I didn't know how many of the people there were ready to hear it. I wasn't at all sure they were ready to see the world through the eyes of an outsider far less attuned to their small, short-term gains than to the planet's huge, long-term loss.

On the way to present my talk, I read a message sent to me by one of the organizers of my event, which was held in the beautiful Clay Center in downtown Charleston. It cautioned me that "President Obama's policies have hit the area very hard, so any talk of climate change would not be well received, unfortunately." I looked up and remarked to myself—actually to the documentary camera crew that was riding along with me—"Well, that *is* unfortunate." I decided not to take that advice. Over that year, the crew followed me from West Virginia to a few college campuses and on to Greenland.

I arrived at my hotel prior to my talk with all the good and bad arguments about coal on my mind. Coal has a noble past powering much of our modern industry, but I strongly believe the "Age of Fossil Fuel" is coming to a close. It has to; it's neither sustainable nor responsible—not for the miners, whose jobs will soon be gone regardless, and not for the rest of us, who have to deal with the emissions. Everybody needs to hear that message, especially those who are most directly affected by it. I planned to pitch to the West Virginians that having an economy based on coal was not a good plan even in the interim, let alone 5 decades down the road. As the film crew and I worked on some Kickstarter merchandise in a conference room, we noticed three guys walking around the hotel. They turned out to be none other than Donald Blankenship and his two bodyguards. Blankenship is the behated [*sic*] "King of Coal," who was awaiting trial for fraud—serious fraud.

Blankenship was notorious in the area. He was a vigorous promoter of mountaintop removal. His effort to undercut his rivals with shady real-estate deals led to lawsuits and fines. He wrote angry memos to his employees: "If any of you have been asked by your group presidents, your supervisors, engineers, or anyone else to do anything other than run coal, you need to ignore them and run coal. This memo is necessary only because we seem to not understand that coal pays the bills . . ." Blankenship also wrote that miners should not worry much about "overcasts," the improvised air ducts that miners carve into the rock and coal to allow potentially explosive methane fumes to escape. Blankenship said these "ventilation issues" would be dealt with later.

Before "later" arrived, there was a huge explosion at the Upper Big Branch Mine that killed 29 miners in a few minutes, some of them over a mile away from the center of the blast. By all accounts, it was easier to convict Blankenship on fraud and racketeering rather than the mean-spirited, thoughtless, and what became fatal routine safety violations. Before he was finally convicted of misdemeanor conspiracy to violate mine safety standards and given a 1-year jail sentence, he was walking around my hotel with his bodyguards and giving everyone the creeps.

Meanwhile, I prepared slides depicting the deceits spread by climate-change deniers funded by the fossil fuel industry and other slides showing the remarkable opportunities for West Virginians if they embraced wind and solar energy. All sorts of new sustainable jobs would be created. The energy would be generated locally, with minimal environmental impact. The water and air would be cleaner and healthier. And the economy would be more stable. Unlike the situation with coal, wind and sunshine cannot be driven out of business by cheaper wind and sunshine from other countries. Renewable energy is produced right where it's used, in West Virginia for West Virginians. The state's citizens could embrace their independent spirit even more than ever. But I wondered about those generations of West Virginians who had made their living from coal. Would the audience embrace a pointy-headed outsider with a message of radical change? Could my message resonate even here?

If you want to change minds, you have to be ready to speak with people who don't see things the way you do. I plunged ahead. Well, about 5 minutes into my talk, I realized that the folks sitting in the Clay Center were pretty much sick of the coal industry. As a longtime semiprofessional comedian, or so they tell me, I could tell by the timing and loudness of the laughs that the audience was with me. Now these were self-selecting fans, a group of West Virginians who decided to spend some of their hard-earned disposable income to see me. But I was in the same town where people had experienced some of the worst effects of coal mining. I went on to show that there were now just 30,000 coal jobs in the whole state, according to the West Virginia

Office of Miners' Health, Safety and Training. That's less than 3 percent of the population in the state. What I'm saying is this: Coal is big business in West Virginia, but it's not *that* big, not anymore. I am sure West Virginians can be happier with a less environmentally destructive, short-term-thinking way of life.

If you're a West Virginian, your National Football League team is usually the Pittsburgh Steelers, because the stadium is right up Highway 79. On any game day, there are more than twice as many people in the stands at any NFL game as there are employed in the West Virginia coal industry. Speaking of Pittsburgh and the Steelers, the team is named for its hometown's traditional industry. Today Pittsburgh's economy is utterly transformed. It still has a little bit of steel business, but it has far more dollars cycling through health care, insurance, and communication. Your mobile phone might find its design and billing system directed from Pittsburgh. Buildings that were once black with coal soot have been scrubbed clean. Industrial land along the waterfront has been remade into a thriving retail district. Change is possible.

As I spoke, I could see I was getting the audience on my side. They not only laughed at my jokes, they saw my point and they embraced my bigger ideas. They realized they could do without this industry of coal and gas extraction that has led to the destruction of the once beautiful West Virginian landscape. I spoke with some of them afterward. Yes, this was a Bill Nye audience, but it was also full of locals who were simply curious to hear what I had to say. I didn't need to bully them about my way of looking at the world. They were ready to hear it. They were already thinking about the world beyond West Virginia and about a future beyond the next paycheck. I tried to offer them an optimistic vision of what that future could look like. Maybe someday soon, West Virginians will lead the world with their transformation to a clean-energy economy. Maybe they can become active agents of change. Here's hoping.

My West Virginia trip got me thinking once again about my grand-father and my uncle Bud. For my grandfather, Uncle Bud's dad, the outdoor life was part of everyday life. In his time, cars were not yet available. Riding a horse was a required skill just for getting around and conducting routine business. It's how you went to work and how you visited your friends. To support the equine transportation sector, there were blacksmiths, stables, mews, livery deliverers, footmen, and saddle shops. That all went away with the arrival of the horseless car-riage. People in the horse business got into another business. It was not good or bad so much as it just *was*. Things change. Jobs change. Designs improve. Nowadays, people ride horses for sport, but early in the 20th century, there was no reasonable hope of preserving the com-mercial horse economy and staving off the automobile economy.

My uncle Bud rode horses, too, but for sport only. He became the "master of the hunt" in the Kansas City area, where he retired. The hunt-ers ride around for a while, then head back to the main stable, where they enjoy a big brunch with orange juice mixed with plenty of cham-pagne. From my grandfather's to my uncle's generation, everything changed in barely 15 years! It's time for another big change—in our energy production.

The documentary film crew and I visited the Boeing plant where I used to work, in Everett, Washington. We met a remarkable guy there. He epitomized to me the spirit of adaptation that we need in West Virginia and all around the globe as we change the way we obtain and use energy. Today, this guy installs flight-critical wire bundles in 747 aircraft. In earlier days, he worked as a stonemason and a bricklayer. He uses his same essential skills—his ability to recognize patterns and to lay materials in place carefully, reliably, and aesthetically—only now he uses those skills to make amazing machines that have transformed the economy and united people from around the world. He was able to adapt to a new set of problems using skills he already had. He was thinking in terms of the big picture and saw that there were new ways to make a living using his traditional skills. He tilted his head at the problem that faced him, and in doing so, he allowed himself to remain open to all the profitable opportunities around him.

So here's the deal, people: In West Virginia and everywhere else, we can see what's happening with our climate, by studying tree rings, pollen counts, ocean sediments, and ice cores. We can see, hear, and smell what happens when we dig up and burn coal. Diligent scientists have filtered the data and evaluated the claims. If we embrace the science, we can see that we need to embrace all the better alternatives, everything all at once. You can help. In many parts of the country, you can now choose to buy carbon-free electricity. You can vote for candidates who favor emissions regulations and fair tax laws for renewable energy and who support the organizations that help promote these constructive ideas. You can volunteer or donate to help the parts of the country (and the world) that are most affected by the energy transition. You can help spread the word, as I did in West Virginia, that the move toward clean energy is a liberation, not an assault.

To be sure, there will be disruptions and there will be pain. Some jobs will no longer be needed or wanted. On the other hand, there will be opportunity and joy. Renewable energy sources offer clean, long-term growth—very much the opposite of loosening regulations to keep dirty coal alive just a little longer. Some amazing new jobs will be created that will be available to people with every imaginable background and training. And if we do nothing, the disruptions and pain will be much, much worse. Earth's climate is changing too quickly for us to adapt without some hard work. The information is all laid out there for us. All we have to do now is the hard part: Make big changes so that we can change the world.

CHAPTER 26

Security Through Nerdiness

Times of crisis bring out the best and worst aspects of the human mind. They activate our tremendous powers of rational problem solving but also unleash potent fears that inspire irrational actions, or no action at all. And make no mistake; we are living in a time of crisis. Managing and getting past it will require all our nerd ingenuity—not just in design and engineering, but also in addressing our emotional needs as people. It's that elusive pursuit of happiness that we want to pursue. It is with all this in mind that I was reflecting on Franklin D. Roosevelt's State of the Union speech of January 6, 1941. War was raging in Europe, and Roosevelt, like most people in the United States, was concerned about national security. The United States had committed to providing the Allied nations with food and other essentials by oceangoing convoy as well as equipping them with some significant combat equipment, including B-17 bombers and crews.

Roosevelt recognized that it might not be long before America got pulled directly into the conflict. He did not want to see that happen, but even more, he wanted to establish a longer-term framework for a world in which such terrible and deadly clashes would no longer exist. In the same way corporate executives direct companies in a top-down fashion, Roosevelt announced a set of principles in his State of the Union speech to guide the United States (and other countries) through the perils of a world at war. If you think of a country as an enormous engineering project, you might call them Roosevelt's principles of design. Or you could adopt the language of politics and simply call it "leadership." The January 6 address has come to be called the "Four Freedoms" speech. In it, President Roosevelt said:

> *In the future days, which we seek to make secure, we look forward to a world founded upon four essential human freedoms. The first is freedom of speech and expression— everywhere in the world. The second is freedom of every person to worship God in his own way—everywhere in the world. The third is freedom from want—which, trans- lated into world terms, means economic understandings, which will secure to every nation a healthy peacetime life for its inhabitants—everywhere in the world. The fourth is freedom from fear—which, translated into world terms, means a world-wide reduction of armaments to such a point and in such a thorough fashion that no nation will be in a position to commit an act of physical aggression against any neighbor—anywhere in the world.*

Those first two freedoms are fairly familiar. Freedom of the press and freedom of worship are institutional rights in the United States, written into the pages of the Constitution by the Founding Nerds, er,

Founding Fathers. Freedom from hunger or want: That one is more of a reach, what my contemporaries call a stretch goal. The United States is a country of great wealth yet poverty exists even here, and extreme poverty persists all around the world, so as a goal, it remains largely unmet. It's fraught with distribution and cultural problems.

But the last one, freedom from fear—to me that is a huge insight. Freedom from fear is what this chapter is all about. I mean, is such a thing possible? Is it even desirable? There are times when fear is useful or essential. It's good to avoid taking steps that lead you off a cliff, for example. There is good reason we humans have such a strong fear response wired into our brains. Nevertheless, Roosevelt was right. We want to be the master of our fear. When fear masters us, the results can be anything from unpleasant to catastrophic.

Mastering fear is key because fear really encapsulates all three of the other freedoms—or rather, the threats to freedom. It emerges from the sense that somebody else may take away something that belongs to us: They might take away our wealth, our power, our religion, our right to say and do what we want. Think about all the anger and conflict in the world that can be traced back to these different types of insecurities, and to the fear they inspire. This "us versus them" view of the world is sometimes known as tribalism, and it is one of the most primal forces in human nature. There is a group (tribe) that we identify with; and then there are all the other groups, which are competing with ours.

In his "Four Freedoms" speech, President Roosevelt never used the word "tribalism," of course (this was a different age), but he clearly got the concept. He saw that when the core freedoms are not satisfied, people compete and fall into conflict; when the freedoms are satisfied, peace can hold. Given the circumstances of the time, he understandably spoke of the path to peace most directly in terms of arms reduction. He was proposing that the governments of the world would be more stable and more secure if their citizens were not on a war footing. Still, he had no illusions that it was enough simply to get rid of the weapons themselves. Achieving peace required addressing the inequalities and injustices that led to the desire for weapons—that is, achieving freedom from fear in all its forms.

This is a very nerdy, design-driven vision of politics, and 76 years later, it still strikes me as a brilliant template for a rational approach to global human rights. Imagine how much more the nations of the world could do if they did not need military spending. Think what we could achieve with a national and international consensus to switch to carbon-free renewable energy and to make it available to everyone in the world. The planet would be cleaner and healthier. Funds now squandered on weaponry could be directed to expand access to food and clean water, to fight disease, to create new technology, to explore the universe. People would live happier and more productive lives. Peace would mean less poverty, and less poverty would mean more peace.

We have the tools we need to get there. The progressive legal structure of the US Constitution is one. Data-filtering to weed out deceptive and inflammatory information is another. Nerd honesty, collective responsibility, methodical implementation—these are all part of the kit, as well. But if we really want a more peaceful world, we have to dig deep into the human psyche and address the feelings of insecurity. Nerds need to go after fear itself.

Fear causes irrational behavior, but fear itself is not irrational. It is an essential survival mechanism that focuses us on the most immediate threats—or rather, the ones we perceive that way. I feel fear that humans won't deal with climate change in time, and I'm one of the most rational people I know (ha?). Tribalism, and the associated fear of people we perceive as different from ourselves, is not inherently irrational, either. We trust the people in our tribe because we know them, or because they are so much like us that we feel like we do. (Here I am defining "tribe" broadly to include those we identify with in culture, class, language, appearance, religion, etc.) Furthermore, if someone in our tribe harms another one of us, we share values and scales of justice or punishment. We know how the rest of the tribe will react.

We often don't trust people from another tribe because they may be unpredictable and potentially dangerous. They might try to steal from us. They might try to kill us. We don't know who they are or what they want, so we have to prepare for the worst. Whatever bad experiences we've had before get projected onto the other tribe because we can't be sure that isn't exactly what they are out to do. It is another form of confirmation bias: We preferentially see trustworthy traits in people who are like us, and untrustworthy ones in those who are not. Nobody is completely immune to it. No, not even me. And not even you, if you are being nerdy-honest.

The us-versus-them mentality is inescapable, whether it is a world war or a family dispute over who gets to keep grandma's quilt. It especially happens between nations, though it can occur at every political level. Be it land, clean water, or beaver pelts, countries always manage to find things to fight about. A popular point of contention is religion. There's a widespread perception that if other people believe something even slightly different from what you believe, there's something wrong with them. "Xenophobia," the fear of outsiders, is from the Greek words for "stranger" and "fear": "Keep all those *other* people out and then we'll have no more trouble" is xenophobia at its powerful worst.

Let me remind us of something. We are all much more alike than we are different. From a biological point of view, we are all almost identical. We share at least 99.9 percent of the same DNA sequences. Within a typical geographic population (such as, say, North American First Nations tribes), the connection is closer to 99.994 percent. A visiting scientist from another star system could hardly tell any of us apart. With this nerdy knowledge, and with world travel becoming more accessible to more of us, the overall trend is that we are all becoming much more accepting of one another. Here's to hoping that happens sooner rather than later.

Much as I love the idea of all the peoples of the world holding hands and singing kumbaya (hold on, that sounds absolutely ghastly, but you know what I mean), I acknowledge that there will always be a bit of tribalism. Boston Red Sox fans will always have issues with New York Yankee fans; as a native Washingtonian and baseball fan, I have issues with both those teams—big issues. One way of dealing with an imperfect world is allowing reasonable and healthy boundaries. We have the old saying "Good fences make good neighbors." Fences mean that you don't have to worry about the family next door—the other tribe—messing around in your lawn and borrowing anything from your garage without asking. Maybe your neighbor doesn't want you to see her sunning herself naked. Maybe she wants to make sure she doesn't see you sunning *yourself* naked. People need privacy. Nations and states need borders to define the scope of their laws.

Mind you, a voluntary boundary is not at all the same as a wall. Keeping to yourself is not the same thing has locking people out. A global lockout is an unworkable idea; even if it weren't, it is backward and contrary to the science of our ancestry and to the concept of equal legal rights for everyone. As nerds, we have to keep our respect for others in play and our fears in check. It relates again to those four freedoms. We know from hard experience that separate-but-equal does not work. It leads to inequalities in wealth and rights, and then we are right back to fear and conflict. Rational thinking rejects a polemical, us-or-them worldview.

And as important as privacy is, it is only one component of freedom from fear. Actively isolating yourself is limiting. It goes against the open exchange of information and ideas that is the hallmark of everything-all-at-once thinking, and much of the modern progress it has produced. In isolation, you will or would miss opportunities for interaction, trade, and the exchange of ideas. I like to think there will come a day when countries do not need to have military forces at all, when warfare is obsolete. I don't think I'll live to see it. I admit, I have a hard time even imaging it. What I can imagine, though, are scientific and technological solutions that create safe barriers and feelings of security on much smaller and less aggressive scales than tanks, cluster bombs, and nuclear missiles. They would bring us just a little closer to President Roosevelt's vision of a world without arms.

When I took over as the CEO of The Planetary Society, I had an opportunity to examine a far-out technology along those lines. I had been invited to speak at a formal dinner at the US Air Force Research Lab in Albuquerque, New Mexico. While I was there, inspecting a wonderful laser stitch-welding process that we use for the sail booms on our *LightSail* spacecraft, my hosts treated me to a demonstration of a new device they were experimenting with: the Active Denial System (ADS). This contraption, er . . . uh, system, drives people away without touching them, without even getting near them. I mean it really, really drives you away—more effectively than a close-talker person with halitosis at a wedding reception.

The ADS features a huge antenna, as large as a king-size bed mounted upright on a station wagon. When it's pointed at you, it hits you with a beam of 95-gigahertz radio waves (that's almost 10^{11} cycles per second and about 10 times the energy in a radar beam). It's like a low-power microwave oven beamed at your entire body. In fact, the total energy in the beam is much lower than the energy in a microwave oven, but the waves are at a much higher frequency that really, uh, gets under your skin. It's the quickness of the sensation that makes it especially, uh . . . effective. Yikes! That day, the ADS antenna was mounted on a Humvee, which made it look extra menacing. The Humvee's diesel engine was driving a large electrical generator, which powered the crazy-powerful ADS setup. And, wow, is it powerful.

The vehicle was parked on a small hill about a kilometer away from where a few of us were standing. Down where I was, the airmen had marked off a rectangular test area, about 4 by 5 meters (13 by 16 feet), with orange traffic cones. They directed me to stand in this box. The instant they swung the antenna and aimed it at where I was standing, it felt like my skin was on fire. My immediate instinct was to run away, to get out of that box as fast as I could. The instant I jumped out of the box (out of the beam area, that is), the sensation was gone, just like that. The idea is to disperse crowds, or to drive people away from some significant place in a contested urban environment or on the battlefield. It's amazing, and a little creepy, how well the ADS works. The truck is obvious to any observer or would-be troublemaker, and the generator is loud. But the beam itself is absolutely invisible.

I have mixed feelings about this particular technology. If armies or police use it in dangerous settings where they previously would have used guns, it would be a step forward. I just don't want to see it being used to break up nonviolent protests and the like. Because this is, so far, military-only technology, there's not a lot of public information available about where and how it has actually been tested. Apparently the military took it to Afghanistan but never used it in combat, maybe because the truck and the antenna are pretty vulnerable. Still, I could see it being more effective and less violent than tear gas and rubber bullets in certain types of conflict areas.

As creepy as it is, the ADS impressed me. It's a product of nerds groping for calmer ways to address violent situations. It uses technology to deal with fear in a less lethal way, which qualifies as a baby step in the direction of easing tribal conflict. Researchers are helping us boost security in many other technological ways: weapons scanners at shipping ports, data systems for passport control, police body cameras and dash-cams, public surveillance networks. They have their own flaws, but at least they are not attack weapons of any form. They all promote freedom from fear, and by extension, they promote the possibility of more rational decision-making—but only in little ways.

I long for the day when all that rational decision-making will allow us to resolve our difficulties with one another for real. I mean, all the way at the source: us. As President Roosevelt laid out in his Four Freedoms, the way to get there isn't by designing better weapons or even by getting rid of weapons entirely. We have to go after all four fears, which means working to reduce tribalism itself. A crisis mentality tends to push people more tightly into their tribal corners. One of the best ways to dial down tribalism and promote peace, then, is to reduce the sense of crisis. Uh . . . how hard can that be?

Oh, right—extremely hard. About the hardest thing there is. We are back to the idea of changing the world. But it's what we are here to do. It's what we are going to do.

I claim that there are opportunities for engineers, scientists, and policy wonks to use technology to make us largely free from fear on a global scale. As the type of person who buys my books (thank you!), you probably know this truth better than anyone: There are two ways to be rich. You can have more, or you can need less. The solution is to do both. Nerd thinking can yield radical improvements in efficiency, but it can also greatly increase the total supply of energy, food, medicine, and information. Perhaps I am being Bill the Wishful Thinker. I could be, in nerdlike fashion, presuming that every problem has, at its core, a technical solution, but I believe it.

Yes, I know tribes of people have been throwing stones and insults at each other for millennia. It's not going to stop tomorrow. Nevertheless, we have to work at resolving conflicts every day. We know the goal, so let's move toward it. I hope that along with our instinct to develop barbed wire, supermax prisons, and Active Denial Systems, we can develop technologies that enable not just tribal-style security but an entire system in which people can trust one another.

I'm thinking of high-efficiency thin-film photovoltaic solar panels, offshore wind turbines, and carbon nanotube power transmission lines so that we can have abundant power for everyone. I'm thinking of new techniques to purify water economically so we don't have to fight over water resources even on a warmer planet. I'm thinking about an across-the-board assault on poverty and scarcity so that there can be radically more freedom and less us-versus-them tribalism all around the world. The correct mix of large-scale and small-scale technologies could seriously level the world's playing field. And if a flat playing field does not satisfy you metaphorically—because our Earth is, after all, round—how about if we work to put all the world's citizens on a sphere of economically equal radius. Is that image too nerdy for you? How about equal opportunity for all? We can do it. Let's get started.

CHAPTER 27

Think Cosmically, Act Globally

I had just turned 13 years old in 1968 when astronaut Bill Anders, on the same *Apollo 8* mission I mentioned back in Chapter 5, took a remarkable photograph of a gibbous Earth with the surface of the Moon in the foreground. That iconic image has come to be called "Earthrise." It was the first time we humans saw our home planet from the vantage of another world. Living here on the Earth's surface, most of us think of the Earth as a huge place with billions of inhabitants living in thousands of cities and millions of villages, but astronauts looking at that picture from far away above the lunar surface see our place in the cosmos differently: a small blue world hanging with no visible means of support—delicate, finite, unique. It is unlike any other planet we have found anywhere in the universe, and it is the only place where we know life can exist (so far).

No one anticipated the profound impact of that Earthrise photograph. Bill Anders looked out the window of his space capsule, saw something amazing, grabbed a camera, and took what he thought was a great-looking view of the Earth. He probably wasn't trying to influence billions of people with a single snapshot. But he did.

As seen from the great height of cislunar space, it is obvious that the Earth is all one world. That alone fills nerds like me with optimism. Before I applied to follow in Anders's footsteps and be a NASA astronaut (unsuccessfully, four times), I was drawn to space and to an optimistic view of the future, perhaps because I grew up with the original *Star Trek*, which aired—when TV was "on the air"—50 years ago. Every week, the starship *Enterprise* (the *NCC-1701*) sailed around the cosmos visiting or battling a different civilization. Each adventure took place not in a country as such, but on an entire world. Since the Earthrise photo came to us, we, too, think of our entire Earth, with all its ecosystems and all its oceans, as one world, just as on that television show—only real. And that changes everything.

The everything-all-at-once nerd worldview is a lot like *Star Trek*'s. Both are rooted in the same philosophy: We are all in this together. The show and its characters have become a part of international culture, like Disney and Mickey Mouse. Everywhere on this planet, you will find *Star Trek* fans. Far more important, everywhere on this planet, you will meet people who were and are influenced by the show and its worldview—or rather, its cosmic view. The show was crafted by Gene Roddenberry, whose stories are based on an optimistic vision of the future in which society provides for everyone. There are no poor people in the United Federation of Planets. The people of the *Trek* universe must have accomplished everything all at once. In their future, every creature comfort is provided through technology. We do not encounter another starship that has run out of food or see a Federation planet where ordinary people are shivering or frozen dead because they lack electrical power. All the routine food, clothing, and shelter problems have been solved. We presume the advanced technology of the future does so much more than we can do now. We further presume that all this comfort is ultimately a product of nothing other than . . . science. After all, these starships have a "science officer." They don't have a "psychic officer," an "alternative medicine doctor," or a "let's all pray harder officer." They have replicators that can take care of practically any need and medical tricorders that can immediately assess any health problem. It's all science all the time in this particular science fiction.

Just like the fictional planets visited by various *Star Trek* starships, Earth seen from the distance of the Moon has no political markings or borders. It also has no natural barriers that confine climate change to one part of the surface. And it is so small compared with the vastness of space. Earth's atmosphere is so thin that if you could somehow drive your car straight up at freeway speeds, you'd be in outer space in less than an hour. The whole idea of "us versus them" looks absurd when you are aware that every person alive is gravitationally held on the same 12,742-kilometer-wide wet rock hurtling through space. There's no option to go it alone. We are all on this ride together.

Unlike in the stories of *Star Trek*, here on our actual planetary ship, we have health problems on global scales. People go hungry. People can't worship as they please. Wars persist, and people cannot live without fear. We have not yet managed to apply our scientific capabilities fully and fairly. We are still far, far from the flat-Earth of global equality as I described at the end of the last chapter. The global poverty rate has fallen by more than half over the past 2 decades, but almost 11 percent of the population (some 770 million people) is still living in a state of extreme poverty, according to the most recent study by the World Bank. The numbers in sub-Saharan Africa have barely dropped at all over that time. And just to be clear, extreme poverty really is "extreme." It is defined as living on less than the equivalent of $1.90 a day. Even allowing for the vagaries of how purchasing power works in different parts of the world, that is a painfully meager existence.

We have to do better. We can do better, if we think big enough. Raising the standard of living for everyone, everywhere all at once, will improve our world and make all of us more secure. When people are able to work and earn a living income, they become productive rather than disenchanted and ready to fight. People raised from poverty contribute more to economic growth and to the expansion of human knowledge. Progress leads to more progress, and we all share in that.

A great many organizations and government agencies are chipping away at the problem of global poverty in nerdy, smart, data-driven ways. I'm a longtime member of the Union of Concerned Scientists (UCS), whose mission statement declares it is "for a safer, healthier world." For example, UCS has engineers who do analysis on vehicle emissions and costs. They make recommendations to the US Congress about achievable fuel-efficiency standards based on their engineering analyses. The organization has experts in agronomy and nutrition, as well, who make recommendations for crop-yield improvement. UCS got its start by raising awareness and sounding alarms with regard to the development, deployment, and maintenance of nuclear weapons and nuclear materials. These are all very meaningful, practical ways to apply nerd thinking to make the world safer and healthier. And although UCS focuses mostly on the United States, it has been influential in protecting tropical rainforests while simultaneously encouraging sustainable development around the world, to make sure that conservation does not come at the cost of the local economy.

Global Citizen (formerly the Global Poverty Project) is an impressive organization that set a goal of ending extreme poverty in the world by 2030. I'm proud to be working with them. Theirs is a difficult goal, but they are going about it in a smart, well-designed way. Global Citizen raises money through high-profile concerts and events as well as through online, targeted fundraising campaigns. The group carefully redistributes those funds, investing broadly in sanitation, education, health care, and financing for small-scale innovations. The Global Citizen programs do not entail just handing out bags of food and then leaving. Experts have shown that such drop-and-run aid programs have a terrible track record. They solve the problem for a day, and then the next, they don't; it's the epitome of short-term action without long-term execution. In contrast, Global Citizen staffers work with local partners and develop long-term, go-the-distance strategies.

What makes organizations like the Union of Concerned Scientists and Global Citizen work well is that they have clear missions that run through all the steps of the upside-down pyramid of design. Start with the idea that you want the world to be healthier. How would you do that? You want people breathing cleaner air and drinking cleaner water. How do you get those things? You can do it top-down or bottom-up. The Union of Concerned Scientists looks at everything all at once and filters for maximum impact. It takes a top-down approach, advocating for regulations that compel people to not pollute. It also takes a bottom-up approach, visiting citizens affected by tide surges and polluted air. Global Citizen solicits contributions from celebrities, corporations, foundations, and individual donors. The funds go to directed efforts such as polio eradication in India and sustainable farming in Africa, another clever mix of top-down and bottom-up.

I strongly believe that taking care of all the people on this planet requires three basic things: reliable and sustainable electricity, clean water, and access to the Internet. Electricity is key because it enables both electronic information flow and energy for water management. One of my favorite organizations leading the renewable-energy effort is the Solutions Project. It's made up of a group of engineers who have done real analyses to demonstrate that renewable-energy sources— wind, sunshine, tides, and geothermal heat—could satisfy the world's entire electricity needs by 2050. The Solutions Project was cofounded by Mark Jacobson, a civil engineer at Stanford University. It raises money to perform the requisite detailed analyses and to pay staffers who advocate for its data-driven proposals in local and national elections.

What makes the Solutions Project stand out is that its proposals are not based on unrealistic intuition, nor do they involve any technologies that do not yet exist. For instance, the Solutions Project engineers propose electrifying all ground transportation: All cars, trucks, and trains would run with batteries or under transmission lines. These technologies exist. I currently (ha?) drive a US-built Chevy Bolt all-electric vehicle. If we had millions of electric cars and trucks fitted with batteries, we could store energy for everyone, everywhere, all the time. I routinely ride between New York and Washington, DC, on the Acela, an all-electric Amtrak train. If the electricity were produced renewably, the trains would run clean. For those who doubt that ideas like this are achievable, please think again. The team at the Solutions Project has. All of its plans are methodically designed, balancing short-term and long-term goals, and based on careful review of the renewable-energy sources that are available near where electricity is needed. Until recently, nobody had done such a methodical analysis of renewable resources; fossil fuels are such a ubiquitous part of our energy infrastructure that it didn't seem necessary. Now that the information and algorithms for updating that infrastructure exist, moving forward will be that much easier.

Over the next 3 decades, in the Solutions Project plan, we would shift from centralized power production to distributed production. It's a big job, but it's a small world. (You can experience both perceptions simultaneously when you contemplate the Earthrise image.) Remember, the energy technologies already exist. They can be shared and configured for local needs everywhere. The Solutions Project people believe we could do all this if we just decided to do it—if we as a society were to just get moving (pun intended). That's why, along with technical work, they advocate for these renewable technologies with local and regional governments as well as with the governments of the United States and more than a hundred other countries. An expanded clean-energy supply will dramatically increase wealth while drastically reducing inequality, tribalism, conflict, and fear. It's an optimistic view of the future through science. It's like *Star Trek* beamed down here to Earth.

I've singled out the organizations that I know best and have worked with directly. I am confident that they do good work in helping to flatten our planet . . . or making it more evenly spherical. But there are many other worthy groups pitching in. With data about charities, as with all information, you want to filter your facts carefully. Look for organizations with relatively low overhead costs (the money they spend just to run their operations) and a proven track record, along with a well-designed mission and set of goals. Watchdog organizations like the BBB Wise Giving Alliance and CharityWatch can give you a quick reality check. They are easy to find online.

❯ ❯ ❯

You may recall that Guadalajara, Mexico, is where the 2016 International Astronautical Congress (IAC) was held, and yes, I was there. It was a wild nerd festival, with vendors selling huge commercial rockets and small-scale student-built spacecraft all under one convention-center roof. It was also a fascinating case study of how the world is starting (but still struggling) to pull together as one. The IAC chose to meet in Guadalajara because it is becoming a leading technology incubator in Mexico. Businesses there have taken a cosmic perspective on the idea of progress. Guadalajara is a pretty well-developed modern town. The economy looks strong, and its facilities are good enough to support a major international conference. But I could see that there is still work to be done. Foreigners like me were strongly encouraged not to drink the water, because our gastric systems don't carry the antibodies that natives have for any number of distress-inducing microbes that are common in Mexican tap water. Technological development has not trickled down to the level of the faucets and the drinking fountains.

This is a city-scale version of a much larger issue: The war on poverty must be waged at all scales, all at once. When you are aiming for the stars, you still need to pay attention to the gutters. I mean that literally. From time to time, a journalist will ask me what I think is the greatest invention by humankind so far. My sense is they're expecting me to say the iPhone or maybe the light bulb. But, I always reply, "Sewers." Without the means to remove human waste and all the sick-making microbes it carries, villages become growth media for all sorts of awful parasites. If the village water that everyone is drinking is contaminated, even slightly, people get sick often. Sick people, kids especially, are unproductive, and they require caretakers. Nearly 800 million people around the world lack clean drinking water, and 2.5 billion do not have access to modern sanitation. Sewers are ground zero if you are targeting clean water for the developing world and, by extension, a large-scale path out of poverty.

In the developed world, civil engineers and municipalities have effectively solved the problem of waste and disease. They provide clean water by use of sophisticated settling tanks, aeration equipment, flocculants (clumping agents), acidity control, and a tremendous system of plumbing. It's so commonplace in Canada, the UK, Denmark, Japan, the United States, and other Western countries that most of us reading along here probably seldom give it a moment's thought. On the other hand, if you grow up without sophisticated plumbing, you probably don't think about it either—because you've never seen it with your own eyes. We in the developed world could fix this. I claim it is in our own best interest to introduce sewers and modern clean-water plants. The national/international service corps I mentioned earlier could be part of the process of making it happen.

Water and energy go hand in hand. Clean water supports a healthy population that can help adapt modern agricultural practices and build a new energy infrastructure. Energy supports the sanitation and purification technologies needed to keep the water clean. Together, they advance the broad cause of bringing more freedom and less fear to the whole world, and promote both fairness and our self-interest. With adequate clean energy and clean water, we can improve health. We can give people access to mobile phones and connections to the Internet. We can expand access to education and economic opportunity. Then we will have more productive populaces, and more productive people to do business with. In the long run, we'll have more nations contributing to the global economy rather than languishing and hoping to receive aid.

As with so many investments, providing modern sanitation and renewable energy for the whole globe looks expensive until you look at the cost of *not* doing it. Just think how much we lose—in productivity, economic growth, science, and technology—by taking so much of the world's population out of the equation. We can all do more to set that right. You can do more. You can donate to the organizations I mentioned earlier, you can volunteer, or perhaps you will start an organization of your own to speed things up. Every nonprofit with a world-changing mission started with a single person. The next person could be you. The progressive goals espoused by the Founding Fathers, and the Four Freedoms articulated by Franklin Roosevelt, are still great templates for global progress and prosperity. When I get revved up, as I seem to be doing now, I feel that failing to embrace these challenges would be downright un-American. Certainly it would be un-nerdy.

Over the past 4 decades, *Star Trek* (along with other idealistic science fiction) has helped seed a vision of a future in which science has solved the world's greatest problems. To pursue that vision, we have to follow the path of everything all at once. *Star Trek* also helped spread awareness—correct, in my humble opinion—that far-out ambitious projects like a space program are not wasteful luxuries, even for developing nations. All the benefits I've been talking (writing) about apply in those nations, as well, perhaps even more so. A space program sets aspirational goals that advance education, innovation, and critical thinking in ways that benefit a whole country, top to bottom, just as NASA's explorations have hugely benefited the United States.

Today, there are fledgling space programs all over the world. India put a probe called *Mangalyaan* ("Mars craft") in orbit around Mars in 2014, and its space agency is enabling the private TeamIndus to land a rover on the Moon in 2018. Countries as varied as Malaysia, Brazil, Iran, Nigeria, and Vietnam have space programs of one sort or another. I'm especially attentive to the case of Mexico, a nation that has been on the sidelines of technological innovation even though it conducts a vast amount of trade with the United States. Mexico's backwater reputation is part of what has made it a target for so many demagogues denouncing immigration and trade treaties. But the situation is changing. Mexico did more than just host a space conference; it has its very own space program now. Guadalajara is even being called (with a little hyperbole, granted) the Silicon Valley of Mexico.

I confess that space is in me, as well as me being part of the cosmos. Carl Sagan started The Planetary Society in 1980. I joined as a charter member. Now I'm the society's CEO. I took the job because I strongly feel that space brings out the best in us. It engages people all over the world. I've argued before that space exploration cuts across political ideologies, but it also cuts across cultures and nations. It knows no tribe, and it is an icon of freedom and the possibility of progress regardless of who you are or where you live. No matter what other troubles humankind has to address, we all want to know where we came from and how we got here on Earth. Our mission at The Planetary Society is explicitly global: We empower the world's citizens to know the cosmos and our place within it. People everywhere yearn to be part of something bigger than themselves.

There are plenty of skeptics who argue that supporting a *LightSail* spacecraft, building a machine to rove on the Moon, or sending a robot to Mars should take a backseat to ending poverty or making the world healthy and safe. They are missing a great big point: When we explore the universe, each of us feels more connected to our world, more fulfilled, and more productive. When *Mangalyaan* reached orbit successfully and began sending back its first pictures of Mars, all of India celebrated. It was a celebration of scientific expertise and good design. It was also a celebration of reason over superstition, collaboration over fear and division. There is not an either-or choice: end poverty or explore space. The two go hand in hand. Exploration and science are part of a larger human drive. I'm proud to be part of it.

I come back to that photograph of the Earth rising above the Moon, a glittering jewel suspended in the void. The impact of seeing our planet from a great distance is so powerful that it has a name: the overview effect. Astronauts who come back from space frequently describe this feeling—a wholesale erasure of the usual sense of boundaries between nations and tribes. We cannot all go into space (yet), but we can all try to internalize that effect and then spread it as widely as possible.

The process of building trust across nations will include sharing knowledge, technology, and the entire rational, nerdy process of problem solving. That is how you strike at the institutional roots of poverty. Those goals are broader than the core priorities of water, energy, and information. They will apply to any of our future needs, as well, and they affirm a commitment to basic human rights. We already have institutions that incorporate the overview effect: not just nonprofit groups like the Union of Concerned Scientists, the Solutions Project, and Global Citizen, but also the United Nations, national service programs, and global businesses. Even the biggest top-down-managed institutions must recognize the need to work bottom-up at the same time, by helping develop local business and local markets.

The scope of the actions can seem daunting, but the overview effect can help us here. We are not all experts at doing these things, nor do we need to be. We just need to work together; we need to respect the principle that everybody knows something you don't. Let's seek out these formal and informal experts and get them on board, or get on board with them. Let's apply the best design principles, testing out different approaches, building prototypes, and refining our solutions.

Let's break the bigger jobs into manageable component tasks. Let's tackle poverty in the best, smartest, nerdiest way, doing our best to accomplish as much as possible—all at once.

Someday soon, I hope, we will see young Americans hard at work here and abroad building wind turbines, installing photovoltaic panels, and running efficient, perhaps revolutionary, transmission lines where they're needed. The next time I visit Guadalajara, maybe Mexico will be launching its own constellation of Internet satellites and everyone from anywhere will be able to drink the tap water without consequence. The next time I look at one of those world-poverty graphs, I want to see all the lines headed down to zero. I dream of the day when we don't need to have these conversations because all these things will really be happening. And I believe that, step by step, we can make that dream real.

CHAPTER 28

Humans Control the Earth; Nerds Should Guide the Humans

There's a term I hear scientists throwing around a lot these days: the "Anthropocene." The word comes from the Greek (it means "the human era") and describes the current geologic era, in which our species dominates the natural processes of our planet. With 7.4 billion people here breathing and burning and dumping into the air, no wonder we are changing the climate. According to one major study, humans have altered 83 percent of Earth's surface, including 98 percent of the areas that are suitable for farming. With our bulldozers, irrigation equipment, and explosives, humans push and shove nearly 100 billion tons of soil a year. We move more earth than the Earth itself does, faster than volcanoes, erosion, and tectonic plates— combined. We are relocating river channels, decapitating mountains, paving grounds and changing drainage patterns with roads, and reworking the composition of the land, sea, and air. We are changing the landscape of the planet at least 100,000 times faster than it could happen in nature. That's how powerful we are.

To me, there's a fascinating and unsettling question raised by the idea of the Anthropocene: When did it begin? I mean, when did humans start to exert a large-scale influence on the Earth? Some people set the date in the mid-20th century, when atomic bombs altered the radioactive composition of the planet's surface. Others set it in the 19th century, with the introduction of chemical fertilizers for intensive farming. Or maybe it started in the 18th century with the Industrial Revolution or in the 16th century when European colonists started transporting species (including infectious diseases) between the Old and New Worlds? I could keep going. I could make a good case that we began throwing around our environmental weight 10,000 years ago with the introduction of agriculture . . . or was it a million years ago with the introduction of fire? I guess I could continue further, but I won't.

The boundary between natural and human influences on the world is a blurry one. We've been changing the Earth in meaningful ways for as long as we've been a technological species. We didn't mean to do these things. Most of the time, we didn't even know we were doing them. People sometimes look at all the modern ecological problems and ask me: How could we, as a species, allow all this to happen? What were we thinking? The fundamental answer is that we *weren't* thinking; we were just doing. We were following our needs and our evolutionary instincts without acknowledging or anticipating consequences on this scale.

This insight could fill a fella with despair. Or . . . you could perform a U-turn and shift your perspective. If you didn't believe me before when I said that we can change the world, maybe you do now, *because we are doing it already*. We cannot turn back the clock, even with the most diligent work by environmentalist do-gooders. We are not able to put removed mountaintops back where they were in West Virginia. We are not going to unbuild cities or uncultivate farms. And we cannot run human progress in reverse—nor should we. Nobody wants to give up comfort, convenience, connectivity, good health, and all the other wonderful things that have come with modern technology. Nobody wants to make the world poorer, less informed, and more tribal. Traditional ways of living are not possible with a modern population. Forward is the only option.

As caretakers of the Earth, we humans have to shape (or reshape) the world the way we want it to be. The only way we can look out for each other is if we also look out for the whole blue ball. We are living as we once imagined beings living in science fiction stories, as masters of our planet. The good news is, we know we have the ability. What we need now is the wisdom, the direction, and the execution to channel that ability in the most constructive ways. Humans are in control of the Earth, but we are not yet fully in charge of ourselves. We need more wisdom, and we need to apply the principles of good design on a global scale. We need experts who have a deep understanding of how our human population interacts with nature. We need to coordinate our actions on a scale never before attempted or even seriously envisioned.

We have started down the right path at least, even if it (yet again) took crisis and conflict to get us started. World War II showed the terrifying possibility of global self-destruction; its aftermath inspired new institutions to promote constructive collaboration on a worldwide-scale. Some of it appeared in the form of international treaties. Some of it appeared as networks of related science, technology, and environmental-research programs. The United Nations, despite its limits and short-comings, provides a forum for international discussion and decision-making. Doctors Without Borders engages physicians from all over to provide medical services to those in need. The Convention on International Trade in Endangered Species (CITES) and Conservation International work to stop poaching and conserve threatened species. The Conference of Parties in Paris in 2015, known as COP21, produced the most meaningful international agreement yet on reducing greenhouse gases.

Ideally, we would have regular meetings establishing regular treaties and agreements akin to COP21 for all manner of global issues. In Bill Nye's perfect world, scientists and engineers from many nations would meet to establish clean air, clean water, and renewable electricity standards. The wealthier nations would work to help the poorer nations by providing know-how and education for all students in every country. We would address strategies for equitable development and poverty reduction. We would collaborate on inspirational projects to cure cancer in all its many forms, on physics experiments, and up and out there in space exploration. These are big ideas, but I often hearken to the incontrovertible truth: The longest journey begins with but a single step.

And don't feel left out, because a lot of those little steps need to come from you. Personal behavior counts. Every penny in a bank adds up. We make decisions about what kind of car to buy, what kind of home to live in, what type of furnace or air conditioner to install. In many places, you can choose how your electricity is produced from the local utility. You have tremendous influence over what your children learn and what kind of values they adopt. You can be an agent of change at your local school. Think about all the ways you can amplify your control over the Earth. International treaties and trade agreements are instrumental in the conservation, preservation, and improvement of the environment. These agreements do not come out of nowhere. They emerge in response to politicians' perceptions of the things that are important in the world—especially what is important to voters. The same is true all the way down through the national, state, and local levels.

As I say every chance I get: Vote. Vote! Wait—maybe you thought I was kidding: *Vote!* That is the fundamental way we influence policies in democratic nations around the world. Right now, it is clearer than ever that elections matter. Even nondemocratic nations are greatly influenced by what they see other major countries doing. Those of us fortunate enough to live in the United States or one of the other progressive democracies around the world have a responsibility to ourselves, and to the rest of the world, to set a good, proactive example. Making your political voice heard is essential because the business of politics is not only setting policies but also finding and choosing our leaders. With the right policies, the right budgets, and the right leaders in place, we can get down to business. Don't get demoralized when an election doesn't go your way. Don't get complacent when it does. Apathy is one of the greatest obstacles to nerdy progress.

Okay, Bill, let's say voting is the most important thing a citizen can do. Then how do citizens choose for whom to vote? This is just another filtering problem. We have to count on advocacy organizations, the ones with a track record of supporting good causes and producing real results. Some of the organizations I mentioned earlier, such as the Union of Concerned Scientists, also pay close attention to which politicians are most supportive and effective in advancing global environmental and human-welfare policies. UCS also conducts independent analyses of the health consequences of public policies and issues guidelines about development strategies to members of Congress and to the congressional staffers who craft much of the policy language. My beloved employer, The Planetary Society, reports on relevant political developments. As it is a nonprofit corporation, we do not lobby in the technical legal sense. Instead, we advocate. We convene expert meetings, meet with elected officials, and encourage our members to petition their representatives. It may sound subtle, but it absolutely is effective. After all, governments are composed of people. Policies are crafted largely as a result of reasoned arguments that are accepted or rejected based on relationships between people.

It's up to all of us to be a part of that relationship, to find those organizations and evaluate their work, tell our friends and colleagues about them, and, ultimately, support them with our money and our time. The complexity of our society requires us to take an analytic approach to voting and the management of our resources. What I'm saying is: We have to all pay attention to politics and be participants in the process, not just kibitzers. Sometimes that means speaking up (calling your local representative, showing up at a town hall meeting) when you don't feel like it. Sometimes that means supporting a politician or piece of legislation even if it (or she or he) is less than perfect. Since you are responsible for a whole planet, there are times when you need to be pragmatic.

As you may have picked up by now, I feel that the more nerdy thinkers we have in high places, the more successful we will be in managing this planet of ours. Over the past few years, there's been a lot of popular backlash against the idea of trusting "experts." I'd find it troubling if it weren't so absurd. Think about it for just a moment. Would air traffic be better without any experts in radar? How healthy would we all be if no one took an interest in the chemistry and microbiology of wastewater treatment? Would our highways be better with no one who understood how to place and light road signs? These are all specialized skills that are fundamentally technical, and we do not want to be without them. I suspect that the attacks on experts are more about tribalism than about expertise itself. Some people fear that addressing poverty abroad means less wealth at home; they imagine that acting globally means offering less help locally. It is an utterly flawed argument from a logical point of view, but it has a lot of raw emotional power. It taps into the us-versus-them mentality.

I hope that thinking and talking in terms of our control of the whole Earth can help. Nobody wins if the planet loses; more relevant, perhaps, nobody loses when the whole planet wins. We want to make the world more fair (at least I hope you do, too), not just because it is the right thing to do but because it is the best thing to do for all of us. I'm not sure there is anyone who doesn't think fairness is a good thing. But fairness is not a stand-alone value. For fairness to flourish, we need to be free to have open discussions and we need to be honest with one another. Echoing Franklin Roosevelt, we also need the basics for a healthy and happy life. In my role as Uncle Bill, the most serious complaints I hear from my nieces and nephews is "But, that's not fair!" ("My sister got to go on the haunted house ride at Funland. Why can't I?" etc.) I don't think it ever changes. Perceived unfairness is what causes conflict well into adulthood. Fairness is an essential element of a better world.

Controlling the planet requires the right policies, check. It requires the right goals, check. It also requires the right science and engineering: the best inputs for any program of directed global change that we might develop. When we are trying to decide what we want to do to the Earth, we need to know exactly what we *can* do, now or very soon.

We know enough right now to establish more comprehensive

agreements and treaties for the fair allocation of water resources, both within the United States and around the world. We can agree not to overfertilize our farms so that nitrogen-rich effluents don't end up in our rivers, gulfs, and seas, where they cause deadly algae blooms; those algae blooms are killing off large populations of fish. And there is no controversy (except among those whose job it is to create controversy) about the need to stop burning coal to cook, heat, or generate electricity. All these things will help bring our planet back into balance.

We can improve our agricultural practices over the next few decades so that we can grow food for 9 or 10 billion people using less of Earth's arable land than we have now. Genetically modified crops are likely to be an important part of the solution. We can engineer plants to tolerate hotter weather, resist drought, and be sturdier in windstorms. We know that these climate challenges are coming. I can imagine inserting new genes into trees so that they are impervious to invasive species we have transported; this challenge is already here. Then I can imagine a system of international agreements that enable farmers and geneticists to create seeds for improved crops or for bug-stopping trees that can be traded through a worldwide development arrangement. These actions will help compensate for global changes we have already made that cannot be undone.

Earth's entire ecosystem is powered by sunshine, and in our nerdy, reengineered world, a lot of our electricity will come from sunshine, as well. The Sun itself will be our generator. I recently learned about Rayton Solar, a California company whose engineers have found a way to manufacture solar cells that produce much less waste than traditional methods. The cells are also thinner and more efficient. Today, typical solar panels can turn about 20 percent of the solar energy they collect into electricity. If we could reach 50 percent efficiency, it would change the world in a hurry. There are many companies, universities, and government research groups working to get there. Then, acting purely on economic interests, power companies would rapidly switch from fossil fuels to solar. Developing countries could set up low-cost solar generators to plug the many holes in their grids. It would become much easier to move toward an economy that no longer enriches Earth's atmosphere with greenhouse gases.

In similar fashion, we can reengineer the landscape so that we pull energy out of the wind as it blows by. We could place wind turbines throughout the Midwest United States. All along the eastern seaboard, where half the people in the country live, there are enormous untapped wind resources. The energy is there; we just have to capture it, distribute it, and find better ways to store it. There is work to be done. We need to reengineer the electric grid to handle the variability of wind power. We need more efficient ways to move electricity around and more cost-effective ways to hold onto it until we need it. These are manageable challenges if we devote enough resources to them and if we approach them in a smart, everything-all-at-once fashion.

These are some of the major tools at our disposal in the Anthropocene. The science and technology that will allow us to do them will come from the usual place: the nerds. The decision to implement them will come from all of us—nerds, nerd sympathizers, and nerds-to-be (I'm an optimist). But to make the right decisions, we need to break down the barriers of tribalism and spread the message that we are in control of the Earth. That's why all the actions I've been talking about are vitally important. If you don't set the agenda for the Earth, someone else will. If we don't control the destiny of our planet deliberately, we will just control it blindly and carelessly. Controlling the planet is not a job you can quit.

Accepting that we all now sit on the Board of Directors of Planet

Earth is a huge new responsibility. It is also a jarring change in how we think of our relationship to nature, especially for those of us who have a long involvement with the ecology movement. Instead of "Let nature take its course," we will be spending a lot more time focused on "We're in charge now. Let's figure out how to run this place." In the long term, it suggests a pretty radical possibility. We are already, by trial and error, finding the levers and switches that control the planet's climate. At some point, we might decide that we need to work those controls on purpose to keep the planet's climate as beneficial as possible—not just reducing our inadvertent impacts but introducing new, intentional ones, as well. We accidentally changed the climate one way. Can we manage to change some aspects of it back, in near–science fiction fashion?

This notion of planet-scale manipulation is often called geoengineering, and some scientists and engineers take it very seriously indeed. (I wrote a lot more about this in my book *Unstoppable*. It's a great read. You should pick up a copy.) They have proposed a number of specific experiments, mostly to adjust the clouds, atmospheric particles, or the color of the sea surface so that some extra fraction of the Sun's energy is reflected back into space. So far, these experiments exist only in the laboratory or as computer models. In the real world, small-scale tests might not be able to yield as meaningful results as large-scale tests . . . Well, you want to be really careful when you are playing around with a planet. We are already living with one set of unintended consequences, and we don't want to introduce another or a few more.

Here's hoping that nerd caution and a step-by-step approach, coupled with a few big ideas, will guide our stewardship. And if some form of geoengineering appears reasonable, here's hoping citizens of the globe will exercise caution and trust in the science. In the let's-take-care-of-everyone future, we can imagine an upside-down pyramid of design for the whole Earth. It would be wild . . . as well as appropriate on a global scale. Being in control of the planet is the greatest test our species has ever faced. It will take all of our best minds, our best filtering, and our best planning. More than ever before, we need to overcome suspicion and superstition, and we need to embrace reason and logic. Let's run our planet effectively and compassionately. Let's run the world nerd-style, to make it better for all.

CHAPTER 29
A Reasoner's Manifesto

Late in the spring of 2016, during the escalating madness of that very memorable presidential campaign, I headed to Washington, DC, to speak at the second Reason Rally. This is an event organized by and for people who want scientific thinking to guide our governments and our legislation. The rally is also a community gathering. In the same way that some people have religious meetings, religious programming, and religious holidays, Reason Ralliers like to hang out together, socialize, and exchange ideas with others who share their basic outlook on life. The Reason Rally consists largely (though not entirely) of atheists. For them, there is no supercompetent deity running the show. In this community, final responsibility for human actions lies with us, and nobody or nothing else. It had intriguing echoes of my trip to the first Earth Day 46 years earlier, only this time I was up on stage instead of down in the audience.

Now, I don't generally talk about religion, except when interviewers seek my views or religious people make a direct assault on science. (When a notorious creationist named Ken Ham, who denies virtually everything we know about geology and the natural history of the Earth, challenged me to a debate, I felt compelled to respond; that experience inspired my book *Undeniable*. Another very fine read that makes a great gift. But I digress.) Other than those instances, I am happy to stick with being the Science Guy and the CEO of The Planetary Society. There are many reasons I've held back. Faith is a highly personal thing; conversations about it have a troublesome tendency to produce misunderstandings and conflict. Even a slight misplaced nuance can inflame tribalism rather than soothing it. But in the big-picture perspective of this book, religious belief is impossible to ignore. It influences how many people filter the information around them, how they think about collective responsibility, and how they respond to, well . . . everything. Data can tell you what is happening in the world, but only your internal code can tell what to do with that information—tell you how to change the world for the better, or whether you should bother to try.

To put things in context, I'm absolutely no stranger to religion. I was raised in the Episcopal Church. I served as an acolyte. I carried the cross and performed all the attendant duties with great reverence. And I still enjoy celebrating Newtonmas, the anniversary of Isaac Newton's birthday as reckoned in Britain, December 25, 1642. (Baby Isaac's mom knew it as Christmas.) As a young man, I was serious about trying to understand the Church's answers to questions I had concerning where I and my fellow Earthlings fit into the cosmos. After graduating from engineering school, I sat down every Sunday and read the Bible. I read it from cover to cover twice; it took me about 2 years. I went to a Christian bookstore and bought the maps connecting biblical events to their real or deduced locations in the Middle East. I had a hard time with the biblical heroes who killed one another, offered to sacrifice their sons, recommended stoning their own daughters to death, and so on. Even if I didn't take those stories literally, they still struck me as morally problematic. I had an even harder time with the idea that a deity killed every living thing on Earth except for one incestuous family and their livestock. That cannot be taken literally, either. The same is true of every miracle in the Bible. If they happened exactly as described, the implications for us humans would be pretty grim.

Here's what I mean. Miracles are magic, and in science you simply *cannot* have magic. A miracle is, at best, a shortcut, a just-so explanation for natural phenomena. I recently met three hard-core, very young creationist Jewish scholars in a bar (surprised me, too). They were earnest and curious; they were also (to my further surprise) fans of *The Science Guy* show. Even so, it took only a moment for their side of the discussion to devolve into magical thinking. Their argument went roughly like this: Science keeps coming up with new ideas, but the Bible never changes. Therefore, the Bible is the only source of truth, and everything we observe with science cannot be accurate, or at the very least, it's unreliable. They suggested that God could have included the whole apparent 4.6-billion-year history of our planet when He created the universe 5,700 years ago. But then they admitted (one more surprise) that God could also have created the world yesterday, with all of our *post facto* memories and histories precisely and consistently set in place, if that was what He wanted. My response was, in essence, "So maybe everything is not real? Hmmm. Really? Enjoy your wine, guys."

I get it, though. People want certainty in the world, but they also want a sense of mystery. Science provides a way through that paradox. Every mystery is a path toward more knowledge. That idea excited me in high-school physics class, and the feeling has never gone away. We will never attain total all-knowing knowledge, admittedly, but what we learn will be real and would be equally real to anyone on the planet (or any intelligent being throughout the universe). That is a level of certainty that no religion can offer. Science can elevate us and soothe us in ways that religion cannot even imagine. The scientist starts from the premise that the natural world is knowable—that reality is real. We can use observation, hypothesis, experiment, result, new hypothesis, new experiment, new result, and so on, to learn more about nature and about ourselves. When we observe something that doesn't fit in with existing knowledge, then we have an opportunity to learn something new. A mystery is not an endgame but rather an opening to a deeper level of understanding. And deep understanding is essential if you want to change the world.

I'm reminded of a wonderful quote from Isaac Asimov (another Isaac), one of the all-time great science explainers. He wrote, "The most exciting phrase to hear in science, the one that heralds new discoveries, is not 'Eureka!' but rather 'Hmmm . . . that's funny . . .'" The implication is that a discovery awaits. It's science. This whole system of thought would fall apart if we could not trust the things we see or, more accurately, observe. Whether you call it a miracle or a magic trick, as an explanation for natural phenomena, it is a dead end. It would be something that, by definition, lies completely beyond rational explanation. With a miracle as an explanation for something, there is no rational way to filter information and test hypotheses. In science, the premise is: We have a process that can enable us to use reason to know nature. The key words are "can" and "know."

You might reasonably ask: When we encounter something completely new and unexpected, isn't it arrogant to start out assuming science is always the right way to go? Well, every good mystery demands investigation, sometimes a thorough investigation. Again and again through history, great and seemingly unsolvable mysteries—from the cause of disease to the movements of the planets—have given way to rich, fascinating, verifiable, and clearly correct scientific explanations.

People at the Reason Rally will tell you that so far through human history, every miracle, effect, or even feeling attributed to religion is much more readily and satisfactorily explained by science. Use your critical-thinking ability and flip the question around: Suppose you observe some confusing new thing and decide it's so strange that science can't possibly explain it. The only way you could reach that conclusion is if you believe that you already know everything there is to know about nature, every possible explanation that could ever be found in the world and cosmos we live in, and that you are the first person in history to bump up against science's outer limit. Now, *that* would be arrogant. (Right, wine bar scholars?)

If we ever do come up against a mystery that is not susceptible to the scientific method, that will be a truly thrilling moment. We will know then that we have encountered some heretofore-unknown force or entity capable of suspending the laws of nature. But if we start attributing familiar events to divine action, we run into big, big problems. People more knowledgeable about philosophy and logic than I am can paint you into a thought-corner pretty quick. If there is a deity who responds to prayer, why does bad stuff keep happening to my family and friends? These days, it's common to see an athlete acknowledge a heavenly deity after he makes a good play. Have you ever seen any of those athletes get angry with a deity when they make a bad play? (Even once?) And what about people who have no knowledge of your deity— what becomes of them? If your deity controls all natural law, how can you count on anything? I'm thinking here again about my companions in the bar, Ken Ham, and their flocks. How can you know when the deity is tinkering with your reality and when you can trust that things are happening in a nominal, normal, predictable way? The questions keep coming, and unlike what we can do with the process of science, there are no good acceptable, or accepted, answers.

I want to emphasize here that I generally have no problem with religious faith; my concern is about using faith to hide from the world rather than engage it. Many religious people get a wonderful sense of community and support from like-minded families and friends. They find inspiration in their faiths to be kind, to be generous to their neighbors, to think in terms of the global good. I respect and admire that, but I long ago realized that the religious approach is not for me. I find my sense of community with my fellow Reason Rally attendees, the ones who get as excited as I do about using our scientific knowledge to make a better world. They are the people I want to hang out with. They are the people I turn to the most for insights on design and filtering. They are the people I see as catalysts for change.

With all this in mind, I thought very carefully about what I wanted to say at the rally. It was quite an honor to be on the stage there, since the other attendees included renowned cosmologists, scientists, and, perhaps most important, magicians (professional foolers of us, foolers for fun). The pressure was on. We were standing in front of the Lincoln Memorial, for cryin' out loud. I put my heart and soul into this thing— in a way that reminds me what we really mean and feel when we say "heart and soul."

The rally took place on June 4, which was not quite summer from an astronomical perspective (the season officially begins with the solstice on June 21), but it sure felt like a Washington, DC, summer-heat kinda day. People largely stayed out of the sun by standing under the trees along the north and south sides of the Washington Monument's reflecting pool. It was an unsettling metaphor for humankind's first reaction to our changing climates: run and hide. But there is no place to run. We have a pressing responsibility right now to be aware, to be bold, and to take charge of our planet. So here is a revised and updated version of what I said to the estimated 15,000 people gathered there. I hope it inspires you to join up with us, or to redouble your efforts, or just to think more deeply about the time we're living in, and put your nerd skills to work.

» » »

I started out like this:

"Ladies and Gentlemen, boys and girls, skeptics, nontheists, and especially the believers who may be here, thank you all for including me in the events today. As we stand before this shrine to Abraham Lincoln, one of history's most thoughtful critical thinkers, I cannot help but feel that we are at a critical time, a turning point in the history of my beloved United States and the history of humankind. Our ability to reason has helped us provide clean water, reliable electricity, and access to an electronic information infrastructure to a large fraction of people in the developed world. Critical thinking, reason, and science got us here. And these traditions will help us bring these technical advantages to everyone on Earth and, dare I say it, change the world." (From here on, I have adapted parts of my speech to include additional thoughts specific to this book.)

Today citizens around the globe are dealing with enormous costs and extraordinary hardships associated with rapidly rising waters and weather events of the extreme kind. We have floods in Texas, terrifying windstorms in the central United States, flooding in southern Germany and Paris, the city that hosted the most recent 21st Conference of Parties (COP21) climate summit. There were over 190 nations in attendance, all hoping to work together to resolve this global-scale problem of atmospheric and oceanic warming—climate change—that has been heretofore largely ignored by most of us in the United States.

Through our industry and agriculture, we have loaded the Earth's atmosphere with carbon dioxide, methane, and other greenhouse gases. We are digging up and burning fossil fuels that are warming our world 10 to the 6th, *one million times*, faster than nature created them. Humankind has brought this on; humankind must address it. Today, our efforts have barely begun. As an engineer and a citizen of the United States, I cannot help but wonder why this is so. Why is this country, which for over a century was the world leader in science, engineering, and innovation, not the world's leader in the renewable-energy technologies, and especially the carbon-curtailing policies, that we must create and put in place as soon as we can?

A handful of climate-change deniers have managed to hoodwink us, to lead us to believe that there is some doubt among the overwhelming majority of scientists about the seriousness and consequences of global climate change, even as our rivers overflow their banks. Without thinking much about it, we allow climate-change deniers to equate routine scientific uncertainty—plus or minus 2 percent, say—with doubt about the observed global changes altogether: plus or minus 100 percent. When I express the situation with these percentages, we all can see that the deniers are obviously wrong or very much misled. Some of the most vocal deniers often suggest that there is a worldwide conspiracy of scientists out to drive coal miners out of work. A conspiracy? Of 30,000 scientists? Have you ever spent any time with these people? They are a competitive bunch. A scientist longs to show that his or her colleagues are wrong. A conspiracy of these people is just not reasonable. Yet a large fraction of us has gone along with the deniers, hardly questioning their inane arguments and obstructionist policy proposals. I have to point out here that nearly half of the country sided with the obstructionists in the last US election. It is troubling, but it may also be galvanizing; it may get us all working together at last.

Climate-change denial is strongly generational. Very few young people embrace those silly ideas. But what kind of future will those kids face? By the time they are old enough to take action, it may be too late. We cannot let them down. We have to meet climate change with change of our own.

The deniers get one thing right: The Earth's climate changes. But the change that is hurtling toward us now is caused by us humans. It's happening fast, and it is not nature working on its own. My grandfather rode into World War I on a horse. It sounds almost unbelievable to many of us now. He was, by all accounts, a skilled horseman. He rode around trenches, in the dark, and under enemy fire. Today, very few soldiers need the skills of horseback riding. The tasks needed to conduct a war have changed. In analogous fashion, a great many jobs will change. People in the extraction industries—those who mine coal or drill for oil and gas—will one day soon be doing something else in the energy sector, welding wind turbine masts, manufacturing photovoltaic systems, or connecting neighbors to the Internet. They won't be out of work. They will be making the future. We can do this.

First, though, we have to get past those who don't believe in a future shaped by science. A consortium of for-profit and not-for-profit organizations, under the aegis of Answers in Genesis, recently opened an amusement park with a religious theme in the commonwealth of Kentucky. You may have heard about its activities. This "ministry," as it refers to itself, preaches that evolution is not real and, more worrisome, insists that our world is not warming. The ministry promotes this fiction among its followers. To work at its Creation Museum or Ark Encounter Bible-literal theme park, you have to testify to your faith in the ministry's faith. That might seem like a violation of our First

Amendment, which you can read from the original text just a few blocks east of here (east of the Lincoln Memorial) in another beautiful building, the National Archives.

To finance these attractions, the Answers in Genesis ministry apparently relies on a consortium of legal entities. It claims religious affiliation when it comes to discrimination in hiring. It relies on Crosswater Canyon, a not-for-profit entity, when it wants to claim the facility will attract tourists to the area, and thereby it should be entitled to tax breaks and virtually free real estate from Kentucky and her taxpayers. While the entities of the consortium have passed legal tests in Kentucky, they did so only because the governor, the tourism cabinet members, and a key judge are all believers. They accept that their religion is not separated from their state or commonwealth, as it would be under other circumstances. I encourage Kentuckians and all of us to imagine if the consortium were about to open something like the Mosque Kiosk, an amusement park or tourist attraction designed to promote the Muslim faith. Do you think that such a project would be defended by the officials who are enabling these biblical businesses to be established? I'm pretty sure the answer would be, roughly, "Hell no."

The story of the Ark Park might seem like an unfortunate but inconsequential, maybe even somewhat quaint, bit of Americana. Nevertheless, there is something very big at stake here: the future. The Ark Project is just one example of a much wider antiscience, antiprogress movement. It is easy to overlook if you live in parts of the country where the influence of this weird worldview is not ubiquitous. But in many other areas, the movement is widespread and persistent. I strongly feel that if parts of this country raise a generation of people trained not to think for themselves, we are all going to pay the price. We are all going to be burdened with reeducating these kids and young adults. The workers in these nearby economies will not have been brought up with the processes of science and reason that help us all understand the world. They will not grow up with the tradition of innovation that has led to the creation of search engines, smartphones, magnetic resonance images, and electric sports cars.

The Ark Park and its associated entities exploit modern technologies as much as any major corporation or business. It's a weird irony. They are using email lists, Web sites, and social media to build a virtual community of believers. They are exploiting the work of scientifically trained engineers and technicians to indoctrinate their flock in ideas that are hostile to scientific and technological progress. And they are appealing to workers in some of the most economically troubled parts of the country, steering them in a direction that will only make their suffering worse and worse. It is frustrating, thoughtless, and heartbreaking.

There is also some positive news here, however. Most religious people do not share a hard-line creationist ideology. Overwhelmingly, people in the United States and around the world want better lives for their children, and they implicitly understand that science and technology are the most effective tools for bringing that about. The Answers in Genesis leaders are trying to escalate that into a broader modern war on science. But, we don't have to play their game. We can expand our fact-based, data-driven movement into something big enough, powerful enough, and appealing enough to overwhelm the forces that don't want or believe in scientific progress.

It is no trouble to find common ground when the environment of the whole Earth is at stake. Pope Francis's recent encyclical "On the Care of Our Common Home" stands out as a commonsense assessment of our planet and its future. It's an example—perhaps the most important example we can readily imagine—of religious people and science-first people finding that they not only have shared goals but also shared methods and even shared philosophies. By any measure, the Pope is an important leader in our world. Just think of all that we can accomplish by working with his people to end extreme poverty, provide education to girls and women, and address as quickly as we can the global need for renewable energy, clean water, and Internet access. As I often say, if you like to worry about things, you're living at a great time. We have suicide bombers, nuclear proliferation, and the Zika virus. But this is also a time of tremendous possibilities and optimism, or it should be.

There will be those who say we don't need to worry because the world is in a deity's hands, not ours. They don't acknowledge how thoroughly the Earth is in our hands now. We just have to disagree with the apostles of inaction and respond forcefully that we take responsibility for our *own* actions. We have to step up and make a better future. That's our Reason Rally way of thinking, but you certainly don't need to be an atheist to embrace it. There are also those who are skeptical of renewable energy and even actively fight to suppress its incorporation into our electrical energy grid. Sometimes they are motivated by old-fashioned financial incentives, as in the case of the West Virginia mining executives: "We sell coal. That's what we do. Why should we stop?" Sometimes they are motivated by a lingering fear that green-energy proponents haven't thought through all the challenges, such as electricity storage and the unreliability of wind and sunshine. They worry, too, that renewables won't be able to supply our needs and sustain our current quality of life.

To those who think we can't get renewable sources in place quickly enough, I give you this response. As I've often mentioned, both my parents were in World War II; their ashes are interred across the river from here (the Lincoln Memorial) in the Arlington National Cemetery.

My father survived nearly 4 years as a prisoner of war captured from Wake Island. My mother was recruited by the US Navy to work deciphering the Nazi Enigma code. They were part of what came to be called the Greatest Generation, but they didn't set out to be great. They just played the hand they were dealt. In barely 5 years, their generation resolved a global conflict and started building a new, democratic, technologically advancing world. With an emphatic sense of purpose, they embraced progress.

The current generation must employ critical thinking and our powers of reason just as they did. This time, the global challenge is climate change. We also must play the hand we have been dealt and get on with it. Together, we can change the world. Thank you.

These days it is all too easy to become isolated within a community of like-minded people, be it at a Reason Rally or political rally. We read each other's Facebook pages, watch the same shows on HBO and Netflix, live in the same types of neighborhoods, visit the same kind of cafés. The technology that we hoped would break down barriers has, in some cases, merely erected new ones. It's in our nature to surround ourselves with like-minded family, friends, and strangers. But we cannot lose perspective of the diversity of opinions, backgrounds, and concerns of those whose lives are unfamiliar. We cannot lose the ability to communicate with those who disagree with us, and we cannot lose our perspective or the ability to identify our common goals.

That is one of our most challenging acts of critical thinking. I will leave it here as a lifelong exercise for the reader. It's a task with many components. Practice at challenging your own reflexive beliefs. Allow yourself to hear opposing arguments honestly and openly. Apply the standard of "prove it" consistently. Respond without snark. Participate in causes where you can work with people united by purpose, not by ideology. We need to be nerds and rally for reason, but outside of the rally, we need to show *why* we rally for reason. Check to see if what you are doing is helping to bring walls down or build them up. Think about what would make a better world. A lot is riding on you.

CHAPTER 30

Design for a Better Future

When I'm in a contemplative frame of mind, I often think about the number 30,000. That is the number of days in a generous human life span: 82 years and 7 weeks, 30,000 shadows circling around a sundial. You probably encounter bigger numbers than that every day. Picture a classic football stadium like the Rose Bowl. If you sat in a different seat every day, you'd only make it a third of the way around. That's what you've got to work with, if you're lucky. It's not a lot of time. When I compare it with the age of the cosmos, the modest span of a human lifetime makes me feel pretty insignificant. The number also reminds me, though, of how much can happen in 30,000 days. Think about how much information and experience you can accumulate, how many people you can learn from and influence, and especially how many actions you can take. When you look at it that way, you have a lot of opportunity to make a big mark on our small planet.

My dad always said he wanted to impart two things to his family: Everyone is responsible for his or her own actions, and you've got to leave the world better than you found it. The first idea is fairly straight-forward. No one else is going to take responsibility for what you do or what you fail to do. If you cheat, it's on you. If you help out, though, you own that, too. The second idea is more complicated. You have to do a lot of work merely to come up with a convincing answer of what "bet-ter" really means. You need to filter information, think critically, con-sider a lot of different viewpoints with honesty and generosity. You need to pay close attention to the design and execution of any projects you undertake. And even then, you cannot be sure about your legacy.

Who hasn't had a desire to peek into the future to see how things turn out? Who hasn't wondered if your life really did make a differ-ence in the world? I think this is what makes time travel such a staple of science fiction, going back at least to H.G. Wells's *The Time Machine* from 1895. It drives the plot of the *Back to the Future* and *The Terminator* movies. As I write, there are four new shows on TV about time travel. Everybody craves a sneak preview of what the rest of our 30,000 days will look like . . . and then the days after those. The disappointing answer from physics is that time travel is impossible. We remember the past, not the future. Information flows one way only. It sucks, I know, but that's the way it is.

We are prisoners of the so-called *arrow of time*, which seems to be related to one of nature's unbreakable rules: the steady increase of entropy, which physicists define as the amount of randomness in a sys-tem. Left on their own, large groups of objects (such as the air molecules in a room) naturally progress from a state of order to disorder. Try this little example: You have a cup of hot tea and a cold ice cube sitting on a saucer. They are separate, nominal objects, not part of the rest of the room. They each hold a different amount of (heat) energy. They are in a state of high order, with all the high-energy molecules tucked in the teacup and the low-energy molecules huddled together in the ice cube. Put them together and their heat energy spreads out into a less independent, intermediate state of lukewarm. If you don't put them together, you still end up with the exact same result. The tea cools to room temperature; the ice melts and warms to room temperature, too. Everything tends toward an in-between, scrambled sameness of energy. The Earth would be run-down if not for the steady flow of energy from the Sun.

Everything we do in trying to change the world is, ultimately, about managing energy and fighting against the tendency toward disorder. That tendency is entropy, and it seems to be a consequence of the nature of time. Or maybe, according to some theorists, the arrow of time is a consequence of entropy. As far as we can tell, the difference doesn't matter to us mortal Earthlings. The bottom line is that we call the direction of time "forward," because as appealing as it would be to be able to turn around, to go back in time and set things right, or set them better, we just cannot. If you feel as though you are constantly struggling against chaos, it's because you are. Entropy is part of what you have to take into account in order to make change happen.

Your body's metabolism is a lifelong battle against entropy. Every cell in your body is doing the same thing you are, trying to influence the future. Your whole existence, everything you do, is a battle against entropy, to make order out of disorder, to make a person out of some chemical compounds. Entropy is not evil, any more than gravity is. The world would not work without entropy. Time would not move without it. We run our planes, trains, automobiles, and electrical grids by means of our understanding of entropy. The famous second law of thermodynamics describes mathematically how entropy works; it also tells us how energy will move through an engine, a chemical reaction, a collapsing magnetic field, or an insulated home. We understand this natural law so well that we are able to mass-produce plastics and effective drugs by means of exquisitely subtle, energy-manipulating chemistry. Through the scientific process, nerdy thinkers have figured out nature's rules and found ways to use them for our benefit.

So entropy not only makes things slow to a stop. It makes us go; I mean it drives us. Consider what your life would be like if somehow you could defeat entropy, if you could reverse that tendency for energy to spread and dissipate. You'd be beating the laws of thermodynamics, and time itself. Ultimately, you could know the future. What would you do differently if someone told you the exact time and circumstances of your death? You couldn't do anything to change that death or else knowledge of the future would not be meaningful. Either you are stuck with a certain fate or the future is not knowable after all. Free will and rational action are unbreakably chained to the unknowable nature of the future. The open-ended possibility of our future days is what sets us free, as analytical nerds and simply as human beings. It's what enables us to be optimistic, to achieve things, to leave the world better than we found it.

Over the years, I've come to realize that most of the time you don't regret what you do; you regret what you *don't* do. How often have you said to yourself, "You know, what I should have done was . . ."? Thinking about the things you'd regret if you didn't do them is a great filtering technique. It helps you think clearly about what you plan to accomplish. That is what made me quit my job at Sundstrand Data Control on a memorable day back in 1986. What are you doing with your freedom? And more to the point, what are you not doing with your freedom that you should be doing? We can't know the future, but we can make the future.

It's this reconciliation with the nature of time and our place on Earth, orbiting a midsize star on the edge of a standard-issue kind of galaxy, that gets me excited about taking my dad's advice. As far as we can tell, humans are the only species on this planet (or in the known universe, for that matter) that can study cause and effect, figure out how to have the most impact, and follow a deliberate strategy to bend the world to our will. We can only play the hand we're dealt. But look! But appreciate it! We *can* play the hand we're dealt. We *can* all do something with our time here. Every day offers a new possibility of action; every moment is precious. We don't fantasize that some deity is in charge, nor do we despair that we are powerless. We know that we can all work together and make change happen. How much more exciting could it possibly be?

Back in the early 1980s, when I worked for a Seattle oil-slick-skimmer company in the oil fields of Texas and New Mexico, I remember the intense, penetrating smell—oil. The petroleum odor would be in your clothes pretty much forever. My coveralls, my socks, even my under-wear smelled like oil. You wash 'em, but the smell stays there all the time. You end up with a set of work clothes that cannot be worn any-where or anytime else. I still have my old coveralls and, all these years later, you can still smell that crude. It is the smell of one kind of future, a stay-the-course approach in which we keep draining Earth's limited fossil fuels and ratcheting up the amount of carbon dioxide in the air.

Recently I returned to Midland, Texas, for a talk at Midland College, in the heart of the oil patch. There, towering above the iconic, nodding grasshopper oil pumps, were wind turbines, enormous freakin' propel-lers in the sky. People who think as I do decided to steer Texas (pun intended) toward a different future, one of renewable clean energy. Those turbines and many others like them are now producing 10 per-cent of the state's electricity. They are beautiful as they spin slowly and powerfully—and there's no smell. The turbines I saw that afternoon are just 20 miles from where I, as a young engineer, programmed oilfield valve controllers 3 decades earlier. In the next 3 decades, oil pumps may be dragged off for scrap steel, melted down, and rebuilt as turbine stanchions. I can't know for sure that's what the future will look like, but I (and you) can help make it, or something very much akin to it, happen.

Everything-all-at-once thinking gives us the tools to identify problems and work out solutions in the most honest, effective way we can. But it lets us do more: Scientific knowledge allows us to make informed projections, so we can anticipate what becomes of those problems and our solutions. This is another way we can fight back against the unknowability of the future. We can come up with very good approximations that let us head off the arrow of time. It is a valuable result of centuries of data collection, critical thinking, and testing of hypotheses using the scientific method: We don't need to blunder all the way into a crisis to know that a crisis is coming.

Now we know, with troubling confidence, that the world is headed for serious problems if we don't address climate change. There's no need to guess about which is better for the planet, more oil pumps or more wind turbines. Greenland ice cores, supercomputer models, satellite observations of Earth, and studies of Venus and other planets all tell the same story. This is a place where we all need to contribute our efforts to make the most of our 30,000 (or more, we may hope) days.

The predictive power of science just astonishes me sometimes. I attended the launch of the *OSIRIS-REx* spacecraft I mentioned earlier. It is traveling to the asteroid Bennu, where it will brush up a sample of 4.6-billion-year-old asteroid dust, tuck the sample into a canister, and fly the canister back to Earth for analysis. The probe launched from Florida's Cape Canaveral on September 8, 2016. After traveling hundreds of millions of kilometers, it will parachute down at the Utah Test and Training Range, 130 kilometers (80 miles) west of Salt Lake City, on September 24, 2023. That is some precision flying.

The launch itself was spectacular and inspiring. The Sun was low in the sky, so when the spacecraft ascended, the smoke trail was beautifully lit. It changed color and character when the solid fuel in the booster rocket was expended. Then there was a spooky shadow from the smoke trail hanging against the darkening blue evening sky. Wow. All that energy, flawlessly harnessed by the scientists and engineers who planned the mission. The technologies developed for *OSIRIS-REx* could someday make it possible to mine resources from asteroids and do manufacturing in space. We are confined to our one planet now, but that may not always be the case. During its encounter with Bennu, the spacecraft will also collect the kind of data we need to know if we ever have to deflect a collision-course asteroid. We may not go the way of the ancient dinosaurs, done in by a flying rock. It is another example of taking fate into our own hands.

And yet there are people—nerdy people, even—who seem a bit uncomfortable with this much free will. When I speak on college campuses these days, students sometimes ask about the "singularity," the supposed moment when computers will be more capable than human brains and the two will merge into some new form of intelligence. An inventor named Ray Kurzweil is the leading proponent of the idea, and it has gained a sizable following, especially around Silicon Valley and the university-rich Boston area. Kurzweil is trying to come up with his own vision of the future, I suppose, but I am disheartened by his notion that people should eagerly await a day when we can surrender control to our devices. It is entirely too mystical and passive for my taste.

More recently, a number of philosophers and computer scientists have been promoting the idea that we are all living in a giant computer simulation. This idea was proposed by an Oxford University philosopher named Nick Bostrom, and even my friend Neil deGrasse Tyson takes it seriously. Um, guys? If some programmer created this world, is it really our fault if it's all falling apart? If we will soon merge with computers that far outclass us in computing power, can't we just let them run the show? When I let myself muse in that direction, I want to stamp my feet and turn around. I hope you do, too. This is dangerously close to the kind of magical thinking that scientists have worked so hard to break away from over the past, oh, 5 centuries.

I encourage people to engage with the world. Join and support organizations that you feel are carrying humankind to a better place. Work together to accomplish great things. Start a movement. Take honest, aware, and fair actions every day.

We have to fight entropy, not just in the chemical reactions that make us living animals on this planet but also in the recurring human impulse to find some way, any way, to shed the feeling of responsibility. Here's where Ned Nye's first piece of wisdom comes into play. *We are responsible.* It's on us to create a better, fairer, healthier world, one in which everyone has access to those three fundamental engineering goals: clean water, renewably produced electricity, and access to the global web of information. We nerds are the right people for this job— rather, these jobs. We're essential. Yes, we're going to do them all, and we're going to do them all at once.

Being a nerd is not easy. It is not just a life of the mind. It offers no escape from the rough, ugly, often unforgiving problems out there. Sometimes those problems are other people who are not interested in facts and are not focused on the greater good. No matter, we still are going to steer into the best future we can envision, guided by the best insights we can glean from the past and from one another.

Each of us carries a unique cache of experience and knowledge with new bits adding on every one of our 30,000ish days. I've shared some of my most memorable experiences with you. I managed to steer the canoe around the rock and haul the "drowning" camp counselor to the dock. I met the creationists head-on and challenged climate-change deniers. I worked over a drawing board, and I resonated in front of a camera. I conceived a cereal-powered car and helped shepherd a *LightSail* into space. I've done my best to learn from every one of these episodes. You have had life experiences every bit as powerful and illuminating as I have had; I'm sure of it. Your nerd duty is to review them. Filter. Look for cause and effect. Keep searching for your own next best steps into the future. Then extrapolate, and expand your views and your reach.

A lot of what I've written about in this book is intended to make a seemingly unmanageable job plausibly manageable. That is a challenge that scientifically minded people have wrestled with for as long as there has been anything that could remotely be called science. We learn about the reality around us by taking in information and by testing ideas. The whole concept of everything all at once is to do that on the largest possible scale and then filter your results with the greatest possible rigor. Without the filtering, the job is beyond human. Even with the filtering, it pushes the limits of our abilities. Critical thinking, open listening, and rigorous honesty are not natural instincts for any of us, at least not in the beginning. They are skills that have to be learned, repeated, and ingrained until they are instincts. Until you steer around magical thinking and tribal impulses as reflexively as I steered around that rock. That's your responsibility.

Then there is the parallel challenge of putting ideas into action. You do not get any points for thinking bold thoughts alone in your room. We all have to be politically engaged. Pay attention to the news and our leaders' views. Find like-minded people; work with them. Find people you disagree with and work to understand them. They know some things that you don't. Reject tribalism. Volunteer for projects and causes you believe in. Entropy constrains our time here on Earth, so find inspiration in that constraint. Nerds need to guide our planet. That's your role in making things better.

There's a third piece of life advice I got from my dad, one that he didn't spell out as clearly as the first two but that has informed my life just as deeply: Treat others with respect. It's the golden rule, for cryin' out loud. This simple principle clarifies the purpose of all the other things we do. It defines the "better" in "make things better." It is what the whole everything-all-at-once technique adds up to. Knowledge has a goal. Scientific theories have an ultimate goal, to explain all of nature. Engineering has a big goal. It all leads to the idea of improving people's lives. Freedom, equality, opportunity, health, peace—they are all facets of the gemstone of human kindness.

With that in mind, we can turn back to the process. We draw on the data and on our personal successes and setbacks to inspire ideas for dealing with the global challenges. If one idea doesn't work out, we keep testing, adjusting, rethinking. We stay aware of those 30,000 days but hold on to our broader perspective of the earlier generations who got us here, and the later ones who are depending on us. We remember the smallness and fragility of our planet, and we relish the opportunity to nurture it. Because of the unforgiving nature of the second law of thermodynamics, we can't know exactly what the future will hold for our descendants. But, we can do our best today to make their future a bright one.

I am at the age now when I know I have more days behind me than ahead. No, no, I'm okay. Strange to say, that doesn't bother me (much). The real gift is being alive here and now, armed with knowledge, free will, and a whole community of like minds. I'm glad you've read this far and come on this journey with me. We inhabit this same moment and have access to all the same tools. All of us, step by step, one by one, working together, can . . . dare I say it . . . change the world.

ACKNOWLEDGMENTS

I'd like to thank the academy . . . wait, that's not what I meant. I mean, thank you Corey Powell! He believed in this book and in me. He convinced me that I really do have something to say that may be of value. Without Corey's continual encouragement, there'd be no *Everything*, let alone *All at Once*. The same goes for Leah Miller, our editor at Rodale, who kept us on track, straightening and rearranging what often started out as tangled ideas. Speaking (or writing) of people who helped form my character and my point of view, we can start with my parents. They raised me in a home with a tradition of academic achievement. They nurtured my love of science with my organic chemist grandfather's glassware, a workbench, a fish tank, various hand tools, and a soldering iron. Susan and Darby, my sister and brother, taught me a lot and pretty much kept me in line. (They're older, and they will always be older.) I couldn't have done much of what I've done over the last few years without the help of my amazing assistant, Christine Sposari, and my patient agent, Nick Pampenella.

I happened to be finishing this book as I was starting a new video show, *Bill Nye Saves the World*. In that writers' room, I worked with some brilliant men and women. Thank you to Mike Drucker, Flora Lichtman, Phil Plait, Abby Plante, CeCe Pleasants, Sanden Toten, Prashanth Venkat, Teagan Wall, and especially Michael Naidus, the show runner and my new good friend. You all brought big ideas and terrific insights.

Finally, I have to thank Jack, John, Gene, Uncle Bob, Uncle Bud, Jeff, and especially George. As you read, you'll meet each of them. They gave me insight and imbued me with my purpose. Now, I want to change the world because I think I can, and because I watched them work. Finally, to the countless fans who have written and spoken to me over the last 30 years and encouraged me to keep at it, thanks to all of you.

Bill Nye, New York, New York
(The town so nice, they named it twice)

My first meeting with Bill Nye took place in a sleepy Brooklyn restaurant, where I asked him why he wanted to write the book that became *Undeniable*. His answer was as heartfelt as it was inevitable: "Because I want to change the world." At that moment, I knew I'd found an ally who shared my optimistic vision of the future. I want to thank Bill for staying true to that vision all the way through to the end of this, our third collaboration. What I appreciate most about him is that he's exactly the same in private as he is in public: smart, curious, funny, and—yes—optimistic, even during the most soul-sucking moments of the writing process.

Looking back, I owe a deep debt to the late Alan Hall, my news editor at *Scientific American*, and to Patti Adcroft, my editorial director at *Discover*, who helped open my eyes to the persuasive power of great writing. And, looking forward, I am grateful to my daughters Eliza and Ava for their daily reminders that a sense of wonder and a sense of social responsibility go hand-in-hand. The goal of this book is to make a better world . . . for them, for all of today's youth, and for all of the generations to come.

Corey S. Powell

INDEX

Underscored page references indicate boxed text. **Boldface** references indicate illustrations.

ABOUT THE AUTHORS

BILL NYE has been the public face of science and discovery for more than 20 years. Best known as the host of the Emmy Award–winning television show *Bill Nye the Science Guy*, Nye is a scientist, engineer, comedian, inventor, and *New York Times* bestselling author of *Undeniable: Evolution and the Science of Creation* and *Unstoppable: Harnessing Science to Change the World.* He is the CEO of The Planetary Society, holds a BS in mechanical engineering from Cornell University, and has seven honorary doctorate degrees.

COREY S. POWELL is the science editor of *Aeon* and the former editor-in-chief of *Discover* and *American Scientist* magazines. He is a visiting scholar at NYU's graduate science-writing program and a contributing editor to such publications as *Popular Science* and *Scientific American.*